The Tallgrass Prairie Center Guide to Prairie Restoration in the Upper Midwest

A BUR OAK BOOK

The Tallgrass Prairie Center Guide to Prairie Restoration in the Upper Midwest

BY DARYL SMITH, DAVE WILLIAMS,

GREG HOUSEAL, AND KIRK HENDERSON

Published for the Tallgrass Prairie Center

by the University of Iowa Press, Iowa City

University of Iowa Press, Iowa City 52242

www.uiowapress.org
Printed in the United States of America

Design by April Leidig-Higgins

The University of Iowa Press is a member of Green Press Initiative and is committed to preserving natural resources.

Printed on acid-free paper

Library of Congress Cataloging-in-Publication Data
The Tallgrass Prairie Center guide to prairie restoration in the Upper Midwest / by Daryl Smith . . . [et al.]. — 1st ed.
p. cm. — (A Bur oak book)
Includes bibliographical references and index.
ISBN-13: 978-1-58729-916-2 (pbk.)
ISBN-10: 1-58729-916-x (pbk.)
ISBN-13: 978-1-58729-952-0 (ebook)
ISBN-10: 1-58729-952-6 (ebook)
1. Prairie restoration — Middle West. 2. Prairie planting — Middle West. 3. Prairies — Middle West — Management. I. Smith, Daryl (Daryl D.), 1938– II. Tallgrass Prairie Center. III. Title: Guide to prairie restoration in the Upper Midwest. IV. Series: Bur oak book.
S621.5.G73T35 2010
639.9'90978 — dc22 2010008379

We dedicate this book to Paul Christiansen
in recognition of his contributions to prairie education
and his ground-breaking work in prairie restoration.

Contents

Acknowledgments

We thank the United States Department of Agriculture Natural Resources Conservation Service for underwriting this guide.

We gratefully acknowledge the support and assistance of our spouses, Sue Smith, Maureen Williams, Ana Houseal, and Terri Henderson; our student assistants, Molly Schlumbohm and Taylor Gerling Shore; the Tallgrass Prairie Center office manager, Mary Weld; our manuscript review committee, Jim Ayen, retired state resource conservationist, Iowa office of the United States Department of Agriculture Natural Resources Conservation Service, Tom Rosburg, professor of biology, Drake University, and Joe Kooiker, roadside biologist, Story County, Iowa; our infinitely patient publisher, Holly Carver; and the University of Northern Iowa.

Finally, our most profound thanks and appreciation to project coordinator and skillful cat herder, David O'Shields.

Why This Manual?

DARYL SMITH

Tallgrass prairie is the most decimated ecosystem in continental North America; less than 2 to 3 percent of the original landscape remains. The blacksoil portion suffered the greatest loss as extensive conversion to cropland between 1830 and 1900 obliterated the prairie, destroying the complex, interwoven fabric of this natural system. Gone from the landscape is the capability for expeditious water adsorption and infiltration and soil formation, grassland animal habitats, diverse collections of interacting organisms, changing panoramas of floral displays, and indigenous cultural lifeways. Though we cannot recreate the original prairie, tallgrass prairie restoration offers the opportunity to actively reverse environmental insult and provide for the recovery of the essential ecosystem aspects of this lost landscape.

The primary focus of this manual is prairie restoration and reconstruction in the tallgrass prairie region of the Upper Midwest, including Iowa, northern Illinois, northwestern Indiana, southwestern Wisconsin, southwestern Minnesota, eastern South Dakota, eastern Nebraska, northwestern Missouri, and northeastern Kansas. Application of information beyond this region is limited as the species composition and vegetation of the tallgrass prairie are affected by climatic variations and soils. For example, to the north, cool-season species increase in importance, and to the south, vegetation is adapted to longer days, milder winters, and an extended growing season. To the west, dry-adapted species become more prominent, and to the east, the prairie intermingles with savanna and woodland. Beyond the designated region, geographic and climate variations necessitate adjustments in the composition of seeding mixtures and modification of dates for planting, seed harvesting, prescribed burning, and other management activities.

This manual is an opportunity for the staff of the Tallgrass Prairie Center at the University of Northern Iowa to share their experience and information with individuals who are restoring and reconstructing tallgrass prairie. The manual is intended to be a useful resource for all types of prairie reconstruction and restoration, including conservation plantings, prairie recovery, prairie recreation, native landscaping in yards, schoolyard prairies, roadside plantings, and pasture renovations. Landowners, conservation agency personnel, ecosystem managers, native-seeding contractors, prairie enthusiasts, teachers, and

Today's prairie remnants are islands surrounded by a sea of agriculture. Photo by David O'Shields.

roadside managers interested in increasing their knowledge and understanding of prairie restoration and reconstruction should find this manual useful. The content level of the manual assumes that the reader has some knowledge of or training in biology and ecology. Individuals with little or no background in biology should collaborate with someone with biological training.

The manual complements two primary publications on tallgrass prairie restoration, *A Practical Guide to Prairie Reconstruction* by Carl Kurtz (2001) and the two editions of *The Tallgrass Restoration Handbook for Prairies, Savannas, and Woodlands* edited by Stephen Packard and Cornelia F. Mutel (1997, 2005). Specific details of prairie reconstruction techniques and procedures in the manual enhance and extend the content of *A Practical Guide to Prairie Reconstruction*. By clarifying the terminology of different approaches to prairie restoration, this manual provides a structural framework for the many topics of *The Tallgrass Restoration Handbook*. In addition, since the only difference in the two editions of this handbook is a new preface, the manual updates information and techniques of restoration gained since 1997. Nebraskans may want to consider the more localized publication, *A Guide to Prairie and Wetland Restoration in Eastern Nebraska* by Gerry Steinauer, Bill Whitney, Krista Adams, Mike Bullerman, and Chris Helzer (2003).

Society benefits environmentally, economically, aesthetically, and culturally

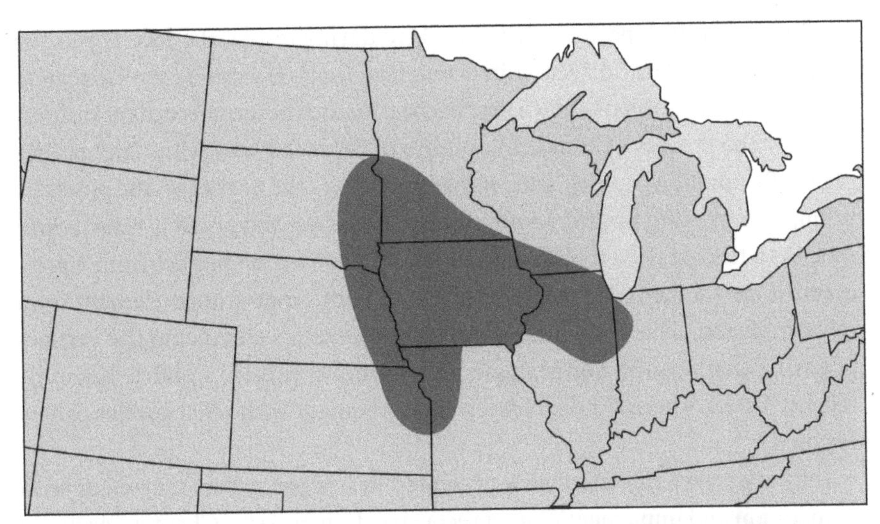

The focus of this manual is primarily the tallgrass prairie of the Upper Midwest. Map by Brent Butler.

from prairie restoration; those involved in this work gain a deeper appreciation and understanding of prairie as well as a sense of pride in their accomplishments. Ultimately, the value of this manual will be measured in terms of the individuals, organizations, and agencies that use it to successfully restore tallgrass prairie to the landscape.

The manual deals with both types of prairie restoration: prairie reconstruction of sites where prairie species no longer exist and prairie remnant restoration of sites retaining relict prairie species.

Terminology regarding prairie restoration has not been consistent over the years. Consequently, there is often a lack of clarity in communications regarding restoration processes and techniques. A description of prairie restoration by Kline and Howells (1987) indicated two basic approaches; one approach creates a prairie community from scratch on a site without existing prairie species, and the other approach upgrades an existing degraded remnant with some relict prairie species present. Previously, the term "prairie restoration" was used indiscriminately to refer to the overall process and to each different approach. The term "restoration" should refer to the overall process; the terms "reconstruction" and "remnant restoration" apply more specifically to each approach.

During the 1990s, we began to use the term "reconstruction" when referring to prairie establishment on sites with no residual prairie species. Since no relict prairie plants are present when the project is initiated, seeds and/or seedlings of prairie species have to be added during reconstruction. This is the process described by Carl Kurtz in *A Practical Guide to Prairie Reconstruction*. To distinguish between the two approaches, we use the term "remnant restoration" for the process of enhancing prairie of degraded sites with relict prairie plants. Restoration of a degraded remnant involves the use of specific practices to improve existing prairie, remove invasive species, and possibly reintroduce species to the site. Key to differentiating a reconstruction project from a project to restore a degraded remnant is the starting-from-scratch element for a reconstruction, whereas remnant restoration involves enhancing the vestiges of prairie that remain. For example, in a former cornfield, a prairie has to be reconstructed, whereas a degraded pasture remnant with relict species can be restored.

Information in this manual is presented in five parts. Part One deals with preparation and implementation of reconstruction projects. Chapter 1 provides an overall summary of the planning and preparation process. Chapter 2 discusses where and how to secure good-quality seed. Chapter 3 discusses designing seed mixes to achieve the type of prairie planned.

Part Two focuses on implementation of the reconstruction plan and early stages of stand establishment. Chapters 4 and 5 describe various methods of preparing and seeding the reconstruction site. Chapter 6 discusses management of the site in its first growing season to aid seedling establishment. Chapter 7 provides information on identifying seedlings and evaluating the success of establishment.

Part Three deals with restoration of degraded remnants and prairie management, including identifying and assessing prairie remnants. Chapter 8 deals with locating, identifying, and distinguishing prairie remnants and assessing their quality. The process of restoring degraded prairie remnants and putting them on a trajectory toward the predetermined restoration goal is outlined in chapter 9. An overall perspective on prairie management, including techniques for managing high-quality prairie remnants as well as restored remnants and completed reconstructions, is provided in chapter 10, along with information about prescribed fire, a primary management technique.

Part Four deals with unique types of prairie reconstructions. Chapter 11 discusses the use of prairie in public spaces. Chapter 12 deals with the use of prairie on roadsides and other erodible sites as part of roadside vegetation management. Chapter 13 considers landscaping on a small scale to develop prairies in yards, outdoor classrooms, and so on.

Part Five focuses on native seed production for those desiring to produce their own seed or wanting to know more about the process. Chapters 14 and

15 describe the processes of harvesting, drying, cleaning, and storing native seed. Chapter 16 discusses techniques and procedures involved in propagating seedlings and transplanting them.

We have included a glossary, a bibliography, and an index. Since, for ease of reading, we have used only common names of prairie species in the chapters, we have also included a list of the common and scientific names of all plants mentioned in this guide.

Tallgrass prairie is a national treasure. As you consider a prairie restoration project, please keep the following in mind. Prairie reconstructions and restorations require a commitment of time, resources, and ongoing management. Progress may be slow, but the processes and product are exciting, fulfilling, and, perhaps, life changing.

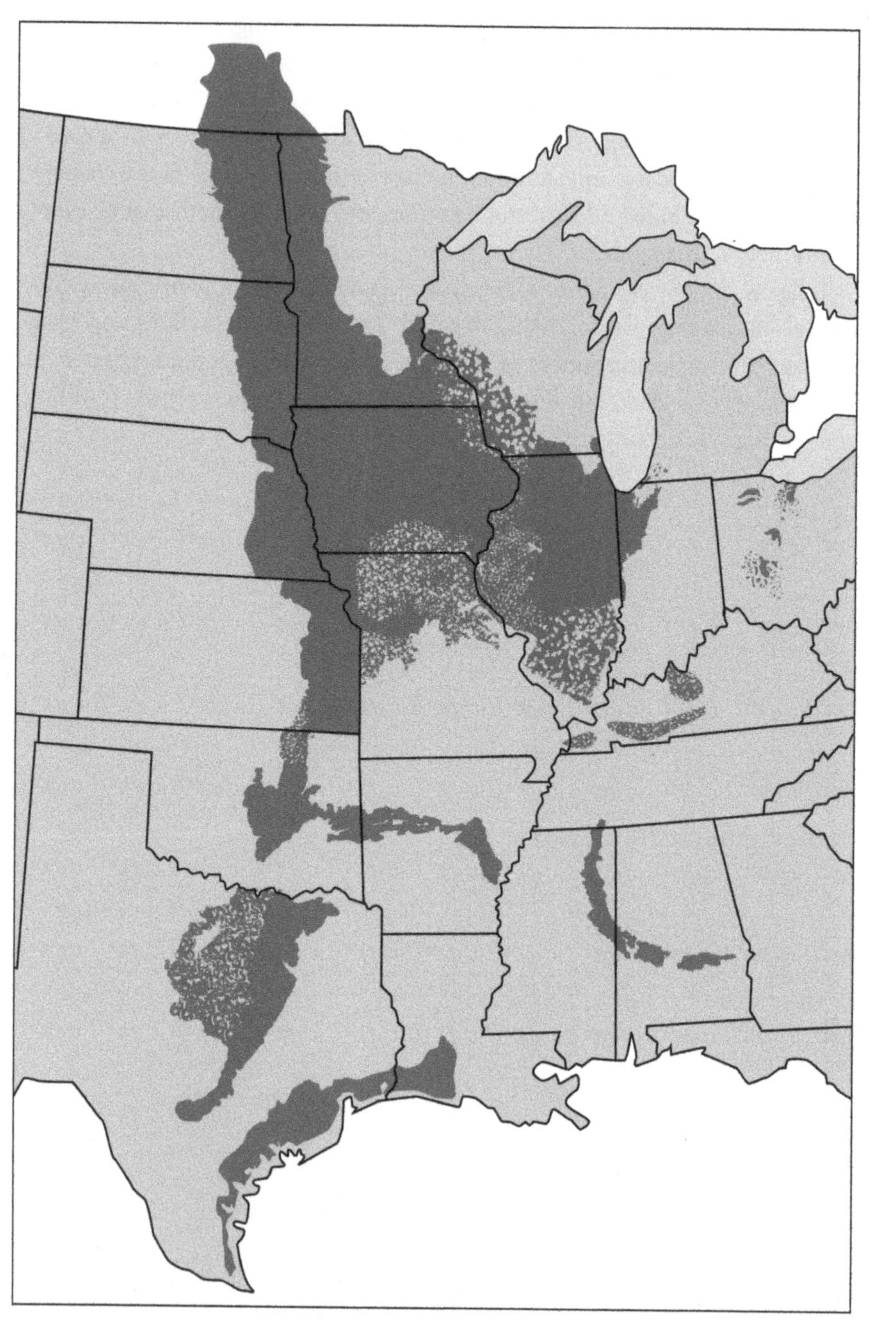

Fig. 1-1. In the early nineteenth century, tallgrass prairie occupied the eastern portion of midcontinental North America between the mixed-grass prairie and the eastern deciduous forest, with a few outliers farther east. Map by Brent Butler.

Introduction: Returning Prairie to the Upper Midwest

DARYL SMITH

The Original Tallgrass Prairie

An immense landscape of grass, wind, and sky once occupied midcontinental North America. This distinctive landscape dominated the horizon from the forest margins of Indiana and Wisconsin to the foothills of the Rocky Mountains and from the Gulf of Mexico to the boreal forest of central Canada. Early French explorers traversing the eastern edge of this wilderness used "prairie," their word for meadow, to describe it.

This vast grassland consisted of three regions distinguished by the vegetation types associated with variations in annual precipitation. The dominant species of the shortgrass prairie occupying the high plains just east of the Rockies were the short, dry-adapted buffalo grass and blue grama. Taller grasses such as big bluestem, Indian grass, and prairie cord grass dominated the tallgrass prairie of the moist eastern zone. Between the two was a shifting zone of mixed prairie distinguished by mid-size grasses like little bluestem, side-oats grama, June grass, and other species representative of both tallgrass and shortgrass prairies. Cool-season grasses were more prominent in the northern prairie. The eastern boundary of the tallgrass prairie interfaced with deciduous forest to form savannas, a mosaic of prairie openings, oak groves, and scattered trees.

Tallgrass prairie encompassed an area westward from the Wabash River to just beyond the Missouri River (fig. I-1). The long grasses dominated most of Iowa and parts of Missouri, Minnesota, Illinois, Indiana, Wisconsin, Kansas, Nebraska, and North and South Dakota. Their dominance extended to the north into southern Manitoba and southward across eastern Oklahoma and Texas to the Gulf Coast. Scattered tallgrass outliers occurred in Ontario, Ohio, Kentucky, Arkansas, and Alabama. In the core of the tallgrass prairie — Iowa, northern Illinois, northwestern Indiana, southwestern Minnesota, and northwestern Missouri — black soil rich in organic matter derived from extensive prairie root systems developed over mineral-rich glacial till.

Tallgrass prairie consists of drought-tolerant grasses, sedges, and wildflowers with occasional shrubs. A good-quality tallgrass remnant may contain up to 300 native species. Grasses dominate by sheer numbers of individual stems and biomass, but the majority of the species are wildflowers and sedges. The aster (daisy) family is represented by more species than any other plant family. The showy flowers of forbs and shrubs intermingle with the grasses and sedges to create a colorful patchwork display. The overall height of the plants of the prairie increases throughout the growing season. Some species bloom briefly; others persist for several weeks. From the first green-up in the spring to the fading russets and golds of autumn, the prairie is an ever-changing panorama of color: purple, blue, lavender, green, yellow, orange, white, pink, and magenta.

The Demise of the Tallgrass Prairie

Tallgrass prairie, part of our biological and cultural heritages, is globally threatened. It is the most decimated ecosystem in continental North America; less than 2 to 3 percent of this original landscape remains. The demise of the tallgrass prairie was rapid and extensive. Within 70 years after Euro-American settlement, by about 1900, more than 90 percent had been converted to agriculture. The blacksoil portion of the tallgrass prairie was hit the hardest; less than 0.05 percent remains. In Iowa, 28 million acres of tallgrass prairie of the presettlement landscape were reduced to widely scattered fragments totaling less than 28,000 acres (Noss et al. 1995; Smith 1998). John Madson succinctly summed the result, "We spent our tallgrass prairie with a prodigal hand, and it probably had to be that way for these are the richest farm soils in the world" (1972).

Most prairie remnants persist on sites poorly suited for row crops. Small remnants exist as isolated islands awash in an agricultural sea, scattered across railroad rights-of-way, roadside ditches, settler cemeteries, and other out-of-the-way places. Occasionally larger pieces (160–200 acres) remain because they were retained for prairie hay. These remnants are small, incomplete samples of the original prairie landscape; as Madson indicates, "To be prairie, really good prairie, it must embrace the horizons" (1972). Some large tallgrass tracts remain on untillable land in the shallow, rocky soil of the Flint Hills of eastern Kansas and the Osage Hills of north-central Oklahoma, the glacial moraines of northeastern South Dakota and southwestern Minnesota, and the Loess Hills of western Iowa.

Conservation organizations such as The Nature Conservancy, natural heritage foundations, and state preserves boards have set aside smaller remnants to preserve tallgrass prairie flora and fauna and convey a limited sense of the original prairie landscape. While these preserves don't encompass the horizon, they are invaluable as refuges for prairie species, excellent study sites, and a nucleus for more extensive prairie restorations. Furthermore, on these remnants, people can reconnect with their biological and cultural heritages.

Prairie Reconstruction and Remnant Restoration

The need for remnant restoration and management emerged late in the history of prairie restoration. Humans drastically altered the tallgrass prairie during the nineteenth and early twentieth centuries. The original prairie was relentlessly whittled down into small, isolated remnants. When prairie preservation began in the middle of the twentieth century, the concept of climax community (Clements 1928) strongly influenced thinking regarding community structure and dynamics. Little thought was given to restoration or management of prairie remnants; managers assumed that remnants would improve in quality as they returned through succession to their prior climax condition. However, acceptance of the Clementsian concept of climax community declined as a more random individualistic concept of community structure and dynamics gained favor (Cawley 1986). In the latter part of the twentieth century, deterioration of prairie remnants made it apparent that ongoing management was essential and that badly degraded remnants would require extensive restoration to recover. The decline in quality resulted from increased exposure to nonnative invasive species, herbicide drift, sedimentation, nutrient overload, and air pollutants as remnants decreased in size, creating a high ratio of edge to interior area. These isolated remnants became increasingly degraded, as critical components of the prairie community disappeared and lost species could not be replaced because extensive cultivation of the surrounding areas had eliminated seed sources.

In the early 1930s, Norman Fassett proposed the first restoration of a prairie (Ritterbusch 2007). This prairie restoration project, initiated in 1935, was part of an extensive program to recreate Wisconsin ecosystems at the University of Wisconsin Arboretum (Sperry 1983). The prairie was later renamed the Curtis Prairie. Aldo Leopold, co-director of the restoration project, recognized that much research and learning could be gained from prairie restoration. At the dedication of the University of Wisconsin Arboretum, he opined, "The time has come for science to busy itself with the earth itself. The first step is to reconstruct a sample of what we had to start with" (Leopold 1934).

Prairie restoration is more than a romantic pursuit or a landscaping novelty.

A restored prairie can provide both environmental and economic benefits. The stems and leaves of prairie plants intercept and adsorb rainfall, reducing the impact on the soil. It is estimated that a big bluestem–dominated prairie will intercept 53 tons of water per acre in an hour-long rainfall of 1 inch (Clark 1937). The extensive root systems are very efficient in absorbing water and increasing water infiltration into the soil, thus reducing both rapid runoff of storm water and soil erosion. Decaying prairie plant roots contribute large quantities of organic matter to the A horizon of the soil and promote soil aggregation. Carbon is sequestered by the root systems that occupy most of the underground prairie, some penetrating 10 feet or more into the soil. Prairie plants are well adapted to low nitrogen levels, and a mature prairie community thrives without fertilizers. With proper management, herbicides are seldom needed to maintain prairie vegetation. Reconstructed prairies and restored remnants provide havens for native birds, insects, and other animals.

Prairie reconstructions differ considerably in purpose, scope, and complexity. They range from relatively small, simplified, stylized reconstructions of natural landscaping in residential areas that create an aesthetic visual essence of native prairies to large reconstructions with more complete community structure and species composition that approximate the original prairie. Reconstructions on corporate sites and public lands can be designed to provide a visually pleasing prairie perspective while controlling storm water runoff and erosion with increased water absorption and infiltration. Reconstructions on former agricultural lands provide the added benefit of increasing wildlife habitat. Larger reconstructions can serve as excellent sites for applied research into the interaction of species, community assemblages, and the development of a functioning ecosystem. Mixtures of prairie species have the potential to be developed as an alternative crop in the production of renewable energy. They can provide carbon-negative, sustainable biomass production while maintaining wildlife habitat. Prairie plantings are also effectively and extensively used on roadsides, federal farm conservation programs, and other locations for reduced maintenance, weed control, improved water quality, habitat development, and enhanced aesthetics. The use of native prairie vegetation on roadsides provides an effective means of weed control as the diverse perennial species fill niches and exclude nonnative species. Furthermore, prairie vegetation on roadsides provides both corridors to connect isolated prairie remnants and pathways for animals to move between sites.

Do No Harm!

Any type of prairie restoration, be it reconstruction or degraded remnant restoration, should be done cautiously. We recommend a conservative approach with "do no harm" as a working guide. Most prairie plants are perennial; mistakes made in a project can persist for a very long time. The following are examples of cautions to consider. When designing seed mixes, note the species that are present in proximal prairie remnants of similar soil and topography. Do not introduce species that are not present in those remnants. When planting prairie adjacent to a native remnant, use seed from the remnant rather than from other locations or a native cultivar. Don't be too quick to plow up or spray out a recent prairie planting if you don't see much evidence of prairie plants. The new seedlings may be slow in establishing, and a good complement of prairie plants may be hidden among the more visible weeds. Allow overgrazed pastures sufficient time to recover after the grazers are removed. Some native species may renew growth quickly after grazing pressure is removed, while other species may require prescribed fire to stimulate growth or to reduce nonnative competition. Before burning, allow time for regrowth of sufficient vegetation to carry a fire. Species of different plants and animals exhibit various responses to fire; to avoid eliminating an entire population, refrain from burning an entire site in the same year.

Perhaps the current diminished condition of the tallgrass prairie is a blessing, a unique opportunity for our society to come together and begin the healing process with prairie restoration. The Ioway believe that the prairie still exists beneath the layers of agriculture and urbanization and that it will emerge if these layers are stripped away (Foster 2004). There may be an element of truth in that belief. Occasionally, when what appears to be a low-diversity, degraded prairie remnant is burned, an awesome and diverse display of wildflowers and grasses appears, almost miraculously, as though they had been awaiting release from a suspended state.

We hope that this manual will aid your efforts to reconstruct or restore portions of the tallgrass prairie and return prairie to the upper midwestern landscape.

Reconstruction Planning

Preparing and Planning for a Reconstruction

1

DARYL SMITH

Overview

Prairie reconstructions should be modeled after remnant prairies. Before beginning a reconstruction project, become familiar with local prairie remnants and available information on prairie reconstruction. A well-designed and well-executed plan is crucial to a successful reconstruction. Components of the plan should include goals and objectives, timeline, budget, site description, designated reference site, description of tasks necessary to complete the project, postplanting management, and monitoring protocol. A successful prairie reconstruction contains many different species of wildflowers, suites of grasses, and sedges that create a panorama of changing colors and increasing heights throughout the growing season.

Preparing for a Reconstruction Project

Reconstructing a prairie is a unique opportunity to return a piece of the tallgrass prairie to the landscape. The original tallgrass prairie was so complex, variable, and diverse that an exact replica is not possible. Prairie reconstruction projects should strive to emulate prairie remnants as nearly as possible within the constraints of budget, site limitations, past land use, adjacent land use, availability of seed, climatic conditions, program requirements, and our limited understanding of the original prairie. Reconstructions can range in size from fifty to a hundred square feet of lawn to tens or hundreds of acres of a conservation planting to thousands of acres in a major landscape reconstruction.

Reconstructing a good-quality prairie requires a long-term commitment. Take time to become knowledgeable about prairie and the process of reconstruction. Read books and other written materials on the topic, examine field guides to become familiar with local prairie plants, peruse native seed and seedling catalogs, visit with people who have had prior experience in reconstruction, and, if available, enroll in workshops on topics related to reconstruction.

If possible, visit local prairie remnants to observe their overall appearance and species composition. They may contain invaluable information for your reconstruction project and serve as a reference site or source of seed. Become familiar with as many species as possible, note where they are located within the prairie, and consider their potential for inclusion in your reconstruction project.

Developing a Reconstruction Plan

Develop a reconstruction plan after you become familiar with local prairies and have an understanding of the process. A successful prairie reconstruction is the result of a well-designed and well-executed plan. The plan will serve as a guide for decision making during the reconstruction process and as a useful reference document for managing the site and future planning. A brief discussion of the components of a reconstruction plan follows.

GOALS AND OBJECTIVES

Prepare a statement of your goals and objectives for the reconstruction project. These goals and objectives will guide your reconstruction activities and determine your final product. Prairie plantings can serve a multitude of functions; however, the intended purpose of the completed prairie reconstruction is the primary consideration for determining your goals and objectives. The following are examples of possible goals for reconstructing different types of prairie: to replicate as nearly as possible the prairie that originally occupied the area, to create a prairie with diverse vegetative composition and structure to enhance native wildlife, to develop a prairie for use by a biology class or student ecology projects, or to develop a prairie to provide forage for grazing cattle. Obviously, the objectives and design considerations will differ considerably for each type of prairie.

Information regarding the historical vegetation of the area will be helpful in preparing goals and objectives. The information can be secured from a variety of sources. Examine the General Land Office (GLO) public land surveys for your county. Surveyor notes and township plat maps of GLO surveys are one of the best, if not the best, sources of information regarding the location of presettlement prairies. A copy of the GLO survey for your county should be available at the courthouse or one of the county offices. In Iowa, the GLO surveys were conducted from 1832 to 1859 as the state was being settled. Paul

Anderson and his colleagues (1996) used the surveyors' field notes and plat maps to produce digitized maps of original vegetation for each county in Iowa. A county soil survey may also provide useful information regarding soils and original vegetation. Additional information to supplement these documents may be obtained from diaries and journals of early prairie travelers, explorers, and settlers; locally written county histories and historical accounts from the middle to late nineteenth century; and recordings of oral histories and family stories passed from generation to generation. Herbarium collections are good sources of information about the past and present location and distribution of prairie plants.

PROJECTED TIMELINE

Develop a timeline with a schedule of the activities required to complete the reconstruction. Knowing the timing of certain activities will insure that materials are available when particular tasks have to be completed. For example, the planned seeding time will dictate when the seed has to be ordered and when the site has to be ready for seeding. If custom operators or volunteers are involved in the process, they will have to be scheduled in advance to be available at the appropriate times.

ESTIMATED BUDGET

A budget is an integral part of a project plan. Cost can be a major factor in determining the size, scope, and quality of the project. Allocate costs for each of the following: seed from commercial growers, herbicide and application or tillage necessary to prepare the site for planting, drilling or broadcasting seed, and postseeding mowing. While not part of the initial reconstruction costs, there will be ongoing expenses for long-term control of weeds and woody species and prescribed burning.

SITE CHARACTERISTICS

Prepare a description of the physical and vegetative characteristics of the site. Information regarding physical characteristics of the site should include topography (slopes, hilltops, and lowlands), soil types, hydrology (drainage patterns, wet and dry areas), amount of exposure to sunlight (nearby shade, aspect of slopes), and plot dimensions.

Topography, soil types, and hydrology of the site are major considerations in determining the type of species to be planted and their location. Certain species do better on higher and drier areas, while others are associated with wetter drainage areas. Some soil publications list native plants associated with particular soil types. Exposure of the reconstructed prairie to sunlight during the day is an important consideration for sites close to buildings or trees. A slope with a southern aspect will receive more direct sunlight than other aspects.

A complete description of the current vegetation of the site is needed to plan for site preparation. Much of the early phase of prairie reconstruction involves achieving an edge over weeds to increase establishment of prairie seedlings. The type of weeds (annual, perennial, and/or invasive) present and the extent of their coverage play a large role in determining how the site will be prepared for seeding.

Maps and photos illustrating site features will also be useful for project planning, development, and monitoring. A sketch map will be sufficient for smaller sites; however, aerial photos, topographic maps, and soil maps will provide more information for larger sites. The photos and maps will also be useful for future maintenance and management activities. Older aerial photographs may not include recent site changes in vegetation or waterways or additions such as cell phone towers or access roads, but they may provide insights into past changes in vegetation. In Iowa, aerial photographs can be downloaded for any site within the state at http:www.ortho.gis.iastate.edu. To obtain an aerial photograph and a soil map, check with your local Natural Resources Conservation Service office or look online at http://www.nrcs.usda.gov/.

THE REFERENCE SITE

The reference site is a model for planning the reconstruction. Since no relict prairie species are present on a reconstruction site, no information will be available regarding the original species composition of the prairie. Select a remnant prairie with soils and topography similar to the reconstruction site. Ideally, one of the remnants visited in the preplanning stage will be sufficiently similar to serve as a reference site. The species composition of the reference site can serve as a model for designing the seed mix to be planted. The reference site can also be used as a comparative standard for monitoring the progress of the reconstruction and evaluating it when completed.

SEED SOURCES

Determine when and where you will obtain seed. Basically there are two options: collecting the seed locally or purchasing it from commercial growers. Good-quality seed can be obtained from a number of commercial growers of native seed. Be certain to order seed well in advance of planting to insure that the grower has a sufficient quantity of seed available. You may elect to hand-collect seed from local remnants during the year prior to planting. This option may be less expensive, and the seed will come from sources known to you. However, seed that appears to be good quality may not germinate. For locally collected seed to be viable, it must be collected at the proper stage of maturity, be properly cleaned, and be stored at low temperature and humidity. You should also determine if pretreatment of certain species is required for germination. For example, stratification, cold treatment, is generally recommended to improve germination, although certain species may not benefit from such treatment. Seed sources and quality are extensively discussed in chapter 2.

DESIGN OF SEED MIXES

Designing a proper seed mix or seed mixes is a key element of a reconstruction, since the species planted are the primary determinants of the species composition of the reconstructed prairie. Objectives for the reconstruction will indicate whether the seed mix is to be mostly grasses, rich in forbs, or an equal mixture of both. Mixes should be varied according to differences in soil types, topography, and hydrology of the site. Reconstructions with a large diversity of species are more attractive than those with a few species and are more representative of the original prairie. Floristic lists for your state will be helpful for selecting species appropriate for your area. The mix design should consider factors such as seed size, viability, competitive ability, availability, and cost. When the design for the mixture is complete, list the species included in the mix and the number of seeds of each species to be planted per square foot. Seed mixes are extensively discussed in chapter 3.

SITE PREPARATION

The primary focus of site preparation is to maximize establishment of the prairie species. Existing vegetation, seeding method, and condition of the soil determine to a large extent the procedures for site preparation. Aggressive perennial weeds should be eliminated prior to seeding. Former crop fields are generally good reconstruction sites, because the weeds have usually been controlled previously and little or no site preparation is necessary. If weeds are localized, spot spraying with herbicide may be sufficient. On the other hand, if weeds are extensive, it may be necessary to till the entire site and/or spray all the vegetation with glyphosate prior to seeding. Site preparation is extensively discussed in chapter 4.

SEEDING TIME AND METHODS

When to seed and the method of seeding are important considerations for germination and establishment of prairie species. The species present in the prairie mix are a primary factor in determining when to plant the seed. For example, a forb-rich mixture will do best when seeded in mid fall, whereas a mix of mostly warm-season grasses will do better when seeded in mid to late spring.

Most reconstructions are either broadcast or drill-seeded, although hydroseeding is used on roadsides and steep slopes. Because special drills are required, a conservation agency or custom operator should be hired to seed larger plantings. Small plots can be seeded by hand-broadcasting.

ESTABLISHMENT MANAGEMENT

In the first growing season after planting, vigorously growing annual weeds usually become abundant while prairie seedlings remain small, utilizing most of their energy for root development. The rapid-growing annual weeds may develop a canopy and shade the small prairie seedlings. Frequent mowing (every 3 to 4 weeks) to a height of approximately 4 inches will curtail weed growth and increase availability of light at ground level for the small prairie plants. With reduced competition and increased light the seedlings will become more robust. The frequency of mowing will vary across the tallgrass prairie; more frequent mowing will be necessary in the moister eastern portion. More vigorous growth of prairie plants with increased available light will improve stand establishment.

MONITORING THE PLANTING

Monitoring will provide information to assess the progress and success of a reconstruction project. The most commonly used parameters are species presence, cover, diversity and density, and biomass production. The actual methods used to collect information are not critical as long as they are consistent and acceptable techniques for sampling vegetation. Monitoring at regular intervals after initial establishment will provide information that will indicate whether midcourse corrections are necessary during the project. Ultimately, monitoring data can be used to evaluate whether the planning goals and objectives have been achieved.

LONG-TERM MANAGEMENT AND PROTECTION

Management of a reconstructed prairie is similar to that of a prairie remnant or restored prairie and is discussed in chapter 10. The management may involve controlling invasive woody species and nonnative perennials such as smooth brome, sweet clover, and Canada thistle. Hand-weeding is an efficient means to control weeds in a small plot, but it requires motivation, an ability to distinguish native and nonnative species, and care to avoid uprooting natives. In larger plots, problem areas can be treated with spot mowing or spot spraying with herbicides. Effective control of woody species will require cutting trees and shrubs and treating their stumps with herbicides. Fire is a very effective prairie management tool to stimulate prairie species, reduce the duff layer, and control woody species. Prescribed fire can be initiated as soon as sufficient vegetation is present to carry a fire. A number of practitioners, particularly in the eastern part of the tallgrass prairie, recommend burning every year during the first decade of a reconstruction to speed up establishment and succession (Schramm 1992).

In developing the plan, give your attention to distinguishing future management units. Management units can be laid out to take into account potential plant communities as determined by topography, hydrology, and soil types. These units can be used for monitoring and applying different management practices such as rotating burns in prescribed fire management.

A plan should include strategies to insure that the completed reconstruction will be protected and properly managed for the future. Adding buffers around the site or obtaining binding protection agreements with adjacent landowners will limit external threats to the integrity of the reconstruction. Conservation easements to maintain reconstructed prairie will provide long-term protection as they become part of the deed for the property. If ownership is transferred to a natural resources agency or organization, a formal agreement to maintain the site should be included in the agreement (Clewell, Rieger, and Munro 2005).

What to Expect from a Reconstruction

A successful tallgrass prairie reconstruction contains many different species of wildflowers that create a changing floral view as different species appear more prominent throughout the late spring, summer, and early fall. The wildflowers are intermixed with shrubs, sedges, and suites of grasses, ranging from shorter, cool-season species that flower and set seed in early summer to taller, warm-season species that become prominent in late summer prior to flowering. The vertical and horizontal structures of the vegetation can vary from a mosaic of patches of shorter plants or clumps of taller ones to widely scattered individuals.

A reconstruction consisting of a diverse, species-rich seed mixture will undergo successional stages and exhibit changes in species composition and species dominance for decades. Although every reconstruction is different, certain species tend to characterize developmental stages. Initial changes in species are rapid and then slow as the planting matures.

The first season is characterized by a mixture of prairie annuals like black-eyed Susan and partridge pea and annual weeds such as foxtail, lambsquarters, and ragweed. Thereafter, annual weeds decline as the native tallgrasses, big bluestem and Indian grass, and aggressive, short-lived perennials like gray-headed coneflower and ox-eye sunflower become more prominent. After a number of years, the aggressive, short-lived perennial forbs decrease in number and longer-lived prairie species like compass plant, the prairie clovers, rough blazing star, rattlesnake master, and spiderwort become more evident. Schramm (1992) asserts that more competition-sensitive, long-lived, slower-growing species, such as leadplant, cream false indigo, and prairie dropseed, become more apparent as the reconstructed community matures.

Species also vary in number and prominence in response to annual fluctuations in climate. These variations are superimposed on the developmental stages inherent in the reconstruction process. Factors adverse to the establishment and persistence of prairie species, such as soil compaction, drought, improper seeding, and/or low-quality seed, may prolong the reconstruction process. Likewise, development may be accelerated with ideal moisture, high-quality seed, and/or a reduction in weed competition. As the reconstruction matures, the vegetation becomes more suitable as habitat for a variety of grassland birds, mammals, and insects.

Summary

- A plan should include the following components: goals and objectives, projected timeline, estimated budget, description of site characteristics, reference site, sources of seed, design of seed mixes, site preparation, seeding time and methods, establishment management, monitoring the program, and long-term management.
- A prairie reconstruction should emulate a local native prairie remnant as closely as possible.
- Visit local prairie remnants to become familiar with the site and vegetation.
- Resources such as books, herbarium collections, original land survey notes, field guides, native seed catalogs, and historical documents can be very useful for practitioners learning about prairie and prairie reconstruction.
- Cost can be a major factor in determining the size, scope, and quality of the project.
- In site selection, consider features of the site that could influence the reconstruction process and anticipate how they may affect the final product.

- A successful prairie reconstruction is the result of a well-designed and well-executed plan.
- A high-quality prairie reconstruction consists of a variety of wildflowers, suites of grasses, and sedges that bloom and set seed at different times throughout the growing season.
- A prairie reconstruction undergoes developmental stages for decades, resulting in changes in species composition and dominance.
- No two reconstructions are the same.
- Provision should be made for long-term management and protection of the completed reconstruction.

Seed Source and Quality

GREG HOUSEAL

Overview

In prairie restoration, seed source and quality are of critical importance. Many seek to use the most appropriate genetic source for restoration. Seed source should not be confused with where the seed is produced or sold; rather, source refers to the original remnant source of the seed. Seed quality is a measure of purity and viability, as tested by a certified seed-testing lab. The following information provides the basis for calculating appropriate seeding rates.

The Importance of Seed Source

In our efforts to restore prairie, we not only look to the remnant prairie as a template for species composition and abundance, we also recognize the genetic composition remaining in prairie remnants and seek to use the most appropriate genetic material. In nature, when soil becomes exposed due to disturbance, plants will recolonize the site from nearby sources. The plant community that eventually develops is a result of viable seed and rootstock remaining in the soil, clonal growth from adjacent plants, and seed deposited on the site by wind, water, and animal activity. In a "pristine" prairie landscape, these recolonizing species will naturally be those appropriate to the soil, hydrology, climate, and fire regime of the site. In a prairie restoration, seed must be selected for the site. This raises at least three important issues regarding seed selection in restorations: selecting species appropriate for the site, choosing appropriate seeding rates for species and combinations of species, and choosing an appropriate source of high-quality seed. In all cases, a seed's source should not be confused with where the seed is produced or sold (that is, the location of a production field, nursery, or seed dealer); rather, source refers to the original remnant source of the seed used to establish the production field or nursery.

Local Ecotype vs. Regional Sources

Prairie restorationists are often admonished to use "local ecotype" seed with little clarification about what this means. As commonly used, both "local" and "ecotype" convey the idea that any remnant population may have specific genetic traits (genotypes) or adaptations unique to that population. Preserving this presumed inherent value of local gene pools of species in remnant communities should be a priority; these scattered fragments are the repositories of genetic material from the original prairie landscape, and from them we can gather seed for future restorations. To help preserve this genetic material, using "local" seed (that is, seed from the remnant itself or other remnants very nearby) is recommended for plantings adjacent to or very near a remnant prairie. Planting adjacent areas with seed from the remnant is an excellent way to buffer the gene pool, as well as buffer other edge disturbances to the plant community. The challenge of this approach is harvesting enough good-quality seed from a remnant in a single year to seed the new planting; therefore, the seeding may need to be done in phases over successive years.

There is an assumption that local seed is always best or better adapted to a proposed reconstruction site than nonlocal seed; however, this may not be true. A single local seed source may be adequate if a large, genetically diverse population is available and seed is collected from throughout the population. On the other hand, extremely small populations of only a few individuals may have lost genetic diversity over time, resulting in decreased reproductive fitness and long-term survival. The subsequent reconstruction from such a source may be compromised over the long term. Knapp points out the possible genetic limitations of using seed from a single source:

> The genetic composition of the individuals colonizing [planted in] a site from a nearby source population usually represents only a small subsample of the overall genetic potential of a species. A newly founded population may establish, persist, and even flourish. But as a result of this limited sampling of the species' overall genetic variation during colonization, the genotypes in the newly established population may be less fit than genotypes available in other populations that, by chance alone, were not introduced into the site. (1996)

In other words, locating a seed-donor site with environmental conditions similar to those of the reconstruction site may be more important than using seed from a local source. For example, harvesting seed from a wet prairie pothole to use on a nearby upland is not as suitable—in terms of both species and ecotypes — as using seed from a more distant but upland remnant source with similar soil and climate and a more appropriate species composition.

A regional source of seed pooled from several remnant populations is an appropriate source for most reconstructions. In the Midwest, remnant prairies are so scattered and isolated that there may be no "local" remnant sources of seed over large areas of the landscape. Various agencies have defined seed-source regions based on geography, landforms, watersheds, species distribution, and political boundaries. Pooling seed collected from different populations in a defined area or region is a reasonable strategy for increasing the genetic diversity of the donor seed, a way of hedging your bet that the right combination of genetic material will be present to best occupy the reconstruction site over the long term. This may be particularly important when donor populations are very small and isolated. Mixing populations from the extremes of a species' geographic or habitat range, however, should be avoided.

If seed nurseries are established as an intermediate step in generating quantities of appropriate seed for restoration, Reinartz advocates using seed from multiple-source populations as foundation seed: "The new genetic population created by combining genotypes of several relict [remnant] populations will form novel genetic combinations, having the potential to evolve entirely new genotypes in a novel habitat. The multiple sources used for establishing the nursery must all be found in the same local area (at least state or region) as the site where the new population will be created" (1997).

An equal amount of seed — or seedling-grown transplants — from each population should be planted in the nursery so that all populations contribute roughly equal amounts to the next generation of seed. Admittedly, in the absence of species- and population-specific information on genetics and reproductive biology, the size and extent of a region and which populations should be mixed often become matters of personal opinion for restoration practitioners.

Bulk Harvest

Native prairie or prairie plantings can be bulk-harvested with a combine, seed stripper, or flail vac. This approach is sometimes used for harvesting seed from a remnant site for broadcast-seeding on adjacent areas to create a buffer zone around the remnant. The quality of the seed and the diversity of species present will depend on the quality and management of the stand. Harvested material is a mixture of seed, chaff, leaves, and stems of species present and in seed at the time of harvest. This material can be analyzed for purity, species composition, and weed seed by certified seed labs, but it is more costly to test this harvested

Fig. 2-1. Sample Seed Test of Butterfly Milkweed Noting Purity, Germination, and Dormancy

Lab. number	12-16-41
Date	1/16/08
Kind and variety	Butterfly milkweed *(Asclepias tuberosa)*
Purity and germination analysis of seed sample	
Pure seed	98.00%
Inert matter	2.00%
Other crop	0.00%
Weed seed	0.00%
Germination	53.00%
Hard seed	%
Total germ and hard seed	%
Dormant seed	27.00%
Tetrazolium chloride	%
Name and no. of noxious weed seed per lb. All state noxious count unless otherwise indicated.	none

material than pure seed of a single species because more time is required for the analysis (fig. 2-1). Material harvested from a well-managed stand may contain 10 to 15 percent seed by weight. This means that a seeding rate of 10 pounds of seed per acre will require 100 to 150 pounds of bulk material to be broadcast per acre. Supplementing bulk-harvested material with seed from very low- or high-growing species, or species that ripen very early or late, is an important consideration since these species may otherwise not be represented in the machine harvest. If purchasing bulk material, request a copy of the seed-test analysis to be sure of the species composition and lack of noxious weeds.

Producers of bulk-harvested seed must take great care to control exotic and invasive species in the stand since they cannot be cleaned out of the material after harvest. Care should be used in cleaning any kind of machinery used in harvesting remnants to avoid contaminating these sites with invasive or non-native species and outside sources of native species. If harvesting from a native prairie remnant, avoid the use of whole-site annual burns, herbicides, fertilizers, or other questionable practices that are detrimental to the long-term health of the remnant prairie communities.

Source-Identified Seed

Because plant materials of known genetic origin were in immediate demand for restoration on disturbed sites in the West, the need for third-party verification of source led to the development of the source-identified seed program. Standards for source-identified, or "yellow tag" seed, were developed by the Association of Official Seed Certifying Agencies (AOSCA). The program is administered by AOSCA's affiliate state crop improvement associations. AOSCA source-identified seed standards provide a fast-track alternative release procedure when there are inadequate existing commercial supplies for a species, propagation material from specific ecotypes is needed for ecosystem restoration, there is a high potential for immediate use, and there is a limited potential for commercial production beyond specific plant community sites (Young 1995).

Source-identified seed for prairie reconstructions may be from either a single remnant or several sources pooled together as a regional source. No intentional selection or testing of traits occurs. Original collection sites are documented, and nursery and production fields established from original collections are inspected and certified annually. Commercially produced seed is marketed with an official AOSCA yellow certification tag, identifying the source and the producer of the material.

As the commercial native seed industry develops, several midwestern states have adopted source-identified seed programs. Individual states differ in their application of source-identified program guidelines regarding native species, so it's important to check specific policies for the particular state in question.

Cultivated Varieties of Native Species

The USDA Plant Materials Program has developed cultivated varieties, commonly known as cultivars, of several native grass and forb species. The traditional approach was to collect an entire plant, or seed from a plant, that exhibited a desired characteristic, such as vigor. These collections, or accessions, were propagated and evaluated in common gardens to further study their characteristics. A selection was made, often of only a few individuals or populations, for further breeding and increase. The goal was to select specific, desired traits, such as good germination and establishment, high forage yield, height, vigor, and winter hardiness. However, while cultivars may be desirable in a pasture setting for forage production, they are not recommended for prairie restoration, particularly since they either have been derived from distant, out-of-state sources or have been selectively bred for specific traits, often competitiveness and vigor, which has the effect of narrowing their genetic base. If cultivars must be used for reconstructions, two or three different varieties should be used to increase the genetic diversity of the planting.

More recently, USDA Plant Materials Center plant selections have reflected the trend toward broad-based regional genetic diversity. Badlands "ecotype" little bluestem, for example, is a composite of 68 accessions (collections) selected for disease resistance from an initial evaluation of 588 vegetative accessions collected from throughout North and South Dakota and Minnesota (USDA-NRCS 1997). This "selection" of a diverse assemblage of little bluestem populations may be a desirable and appropriate seed source for restorations in those states from which it was derived.

The USDA Natural Resources Conservation Service has long recognized the limits of the successful transfer of plant material to another location (Cooper 1957; McMillan 1959). Their recommendation has been to not move cultivars (primarily in reference to warm-season grasses) more than 300 miles north or 200 miles south of their origin. Plants moved northward more than 300 miles generally will not produce seed and are prone to winter injury (Jacobson 1986; Olson 1986). Plants moved southward farther than 200 miles are more prone to disease.

Cultivar material exists only for a few native species. Many native species that are in demand for restoration can be obtained only through direct harvest from native stands or through the source-identified seed program described above.

Criteria for Selecting a Seed Source

Awareness of the types of seed sources available for restoration, along with some planning, will allow you to give careful consideration to the issue of seed source for your prairie planting. Guidelines for recommendations are summarized in table 2-1. Choosing an appropriate source of seed for a restoration planting is a case-by-case determination, most likely requiring a balance or blend of the following realities: proximity to remnant communities that might be negatively affected by introduced genotypes and/or species; the intended purpose of the planting, that is, economic use as native pasture, forage, or erosion control or environmental use for beautification, habitat, or biogenetic diversity; and the budget constraints of the project.

Table 2-1. Prioritizing Seed Source According to Planting Goal

Goal	Local Remnant	Regional (Source-Identified Seed)	Cultivar
Restoration	x		
Reconstruction	x	x	
Forage/biofuels		x	x

In balancing these realities, priority should be given to preserving gene pools present in a remnant prairie. Plantings adjacent to or within half a mile of a remnant prairie should use seed from the remnant itself if sufficient populations and species diversity exist. Keep in mind that a very small or degraded remnant may lack species diversity or may harbor very small populations (only a few individuals) of an appropriate species desirable for the site. Seed of these species from other remnants of similar soil and hydrology in the area may be desirable in these situations. Ideally, the characteristics of the donor site or sites should match those of the reconstruction site as closely as possible.

Regional sources of seed pooled from several remnant sources are an excellent choice for high-quality to moderate-quality reconstructions not affecting local remnants. Regional sources have a broad genetic base that increases the chance that the right genotypes are present to best establish and persist in the reconstructed prairie. Source-identified seed is your best assurance of source, short of collecting the seed yourself or working directly with others you trust. Source-identified seed quality and purity are high, and species can be purchased individually and custom-mixed for specific sites. Many nurseries also provide special pre- or custom-mixed species for wet, dry, and mesic sites. It is a good idea to review the list of included species to be sure they are native to your area and are of acceptable source for your restoration goals. Expect your seed to be delivered with certification tags and seed-test results attached. If not, take it back and request a refund.

Seed Quality

Knowing the quality of the seed that you purchase or plant is critically important to the success of a restoration. Impure or trashy seed will not store as well as clean seed and can harbor the seed of noxious weeds or other undesirable species. Dead seed won't grow; dormant seed won't germinate immediately. Fortunately, seed quality has improved dramatically over the past few years as growers gain experience and acquire better equipment for producing, harvesting, and cleaning native species. Seed dispersal apparatus like awns on grass seed and hairy parachutes on forb seed are routinely removed. This means that the seed lot can be cleaned to greater purity and viability and will flow more efficiently through the seeding equipment.

Native seed is sold on a pure live seed, or PLS, basis. Three factors are used to calculate the percentage of pure live seed: purity, germination, and dormancy. These values must be determined by a certified seed analyst. Purity is a measure of pure, unbroken crop seed units as a percentage by weight of the seed lot. The percentage of germination is determined by placing seed in a germination chamber for an approved time period. Many species, particularly forbs, have dormancy mechanisms that require several weeks of cold, moist stratification to break dormancy, allowing germination to occur. For most native species, no standard protocol exists for breaking dormancy for germination-testing purposes. Therefore, any remaining nongerminated seed is tested biochemically with tetrazolium chloride (TZ), a clear compound that stains living tissue cherry red. The analyst determines the potential viability of stained seed; nongerminated seed considered viable by a TZ test is counted as dormant. A seed test showing a high percentage of dormancy is common in many native forb species, most legumes, and some grasses (table 2-2). This should be expected of natives, particularly in seed lots harvested within the past year.

Calculating Pure Live Seed Amounts

PLS is a measure of the proportion of the viable seed—seed that potentially will germinate—of a species or variety per unit weight for a given lot of seed. PLS for forage crops and turf grass is normally calculated using percentage of purity and percentage of germination only, as dormancy is not a significant issue for these types of species. Native species, however, may have a significant proportion of dormant yet viable seed, particularly among forb species. The native seed trade recognizes this fact and uses all three factors—purity, germination, and dormancy—to calculate the PLS of any given native seed lot. Calculate pounds (#) of pure live seed as follows:

PLS = (bulk pounds) × (% purity) × (% germination + % dormancy)

Where percentage is expressed as a proportion, that is, 98% = 0.98

For example, a 50-pound bulk bag of seed that is 98 percent pure seed, with 53 percent germination and 27 percent dormant seed, really contains only 38 pounds of pure viable seed:

Table 2-2. Purity, Germination, and Dormancy from an Actual Seed Test for the Species Listed

Species	Purity	Germination	Dormancy	PLS
Warm-season grasses				
Big bluestem	90.19	50.0	44.0	84.8
Side-oats grama	89.95	90.0	0.0	81.0
Bluejoint grass	98.37	80.0	5.0	83.6
Canada wild rye	98.31	73.0	20.0	91.4
June grass	95.39	74.0	5.0	75.4
Switchgrass	99.94	40.0	55.0	94.9
Little bluestem	82.53	54.0	33.0	71.8
Indian grass	94.75	28.0	59.0	82.4
Prairie cord grass	85.83	91 (TZ)	—	—
Tall dropseed	99.91	90.5	0.0	90.4
Prairie dropseed	99.22	43.0	6.0	48.6
Porcupine grass	62.89	22.0	57.0	49.7
Legumes				
Leadplant	96.35	7.5	84.0	88.2
Milk vetch	95.77	6.0	86.5	88.6
Cream false indigo				0.0
White wild indigo	99.98	3.5	95.0	98.5
New Jersey tea	92.06	12.0	68.0	73.6
White prairie clover	99.87	77.0	5.0	81.9
Purple prairie clover	99.69	21.0	74.0	94.7
Showy tick trefoil	99.48	26.5	65.5	91.5
Round-headed bush clover	99.74	62.0	34.0	95.8
Wildflowers				
Canada anemone	99.05	21.5	70.5	91.1
Thimbleweed	98.41	91 (TZ)	—	—
White sage	98.95	70.0	13.0	82.1
New England aster	91.75	49.0	24.5	67.4
Smooth blue aster	99.24	71.0	19.0	89.3
Sky-blue aster	99.22	72.0	22.0	93.3
Butterfly milkweed	98.37	50.5	40.0	89.0
Prairie coreopsis	70.30	42.0	47.0	62.6
Pale purple coneflower	93.65	43.5	50.0	87.6
Rattlesnake master	95.78	24.5	71.0	91.5
Bottle gentian	75.83	3.5	82.0	64.8
Ox-eye sunflower	99.84	71.0	19.0	89.9
Rough blazing star	98.41	36.0	61.0	95.5
Prairie blazing star	98.28	3.5	92.5	94.3
Great blue lobelia	92.20	21.5	38.0	54.9

Table 2-2. (*Continued*)

Species	Purity	Germination	Dormancy	PLS
Wild bergamot	95.51	62.5	2.0	61.6
Wild quinine	96.13	81.0	10.0	87.5
Common mountain mint	97.06	77.5	8.0	83.0
Hairy mountain mint	98.70	92.0	0.0	90.8
Slender mountain mint	95.03	75.5	11.0	82.2
Gray-headed coneflower	98.68	89.5	0.0	88.3
Fragrant coneflower	98.02	71.0	25.0	94.1
Rosinweed	81.03	40.0	54.5	76.6
Compass plant	86.88	23.0	72.0	82.5
Stiff goldenrod	95.31	60.0	22.0	78.2
Showy goldenrod	86.45	26.0	66.0	79.5
Prairie spiderwort	99.98	5.0	51.0	56.0
Ohio spiderwort	99.76	6.0	74.0	79.8
Culver's root	93.04	22.0	37.5	55.4
Golden alexanders	99.33	22.0	61.0	82.4

Note: A high percentage of dormancy can be expected in most legumes, many wildflowers, and even some grasses. Expect a high percentage of hard seed in legumes.

PLS = 50# bulk × 0.98 × (0.52 + 0.27) = 38# PLS or 50-pound bulk × 0.7742

If, however, you request a 50-pound PLS bag of that same seed, you would receive a bag weighing 64.58 pounds:

Bulk pounds = # PLS / [(% purity) × (% germination + % dormancy)] or 50# PLS / 0.7742# = 64.58 bulk pounds

Summary

- In prairie restoration, seed source and quality are critically important.
- Source should not be confused with where the seed is produced or sold; rather, it is the original remnant source of the seed harvested or used to establish the production field or nursery.
- Preserving the presumed inherent value of local gene pools of species in remnant communities should be a priority.
- Using seed from the remnant itself or other remnants very nearby is recommended for plantings adjacent to or within half a mile of a remnant prairie.

- The assumption that local seed is always best or better adapted to a proposed reconstruction site may not be true.
- An important consideration is to match the characteristics of the seed-donor site to those of the reconstruction site.
- A regional source of seed pooled from several remnant populations is an appropriate source for most reconstructions.
- Use seed from multiple-source populations to establish new nursery populations as foundation seed for prairie restorations.
- Source-identified seed may be from either a single remnant or several sources pooled together as a regional source.
- Although cultivars are not recommended, the use of two or three different varieties of the same species will help increase the genetic diversity of the planting.
- Bulk harvesting may be appropriate for harvesting seed from a remnant site for use on adjacent areas to create a buffer zone around the remnant.
- Bulk-harvested material is a mixture of seed, chaff, leaves, and stems of species present and in seed at the time of harvest and may contain 10 to 15 percent seed if well managed.
- The quality of seed and the diversity of species in bulk-harvested material will depend on the quality and management of the stand.
- If purchasing bulk material, request a copy of the seed-test analysis to be sure of species composition and the lack of noxious weeds.
- Take care when cleaning any kind of machinery used in remnants to avoid contaminating these sites with invasive or nonnative species or alien sources of native species.
- A high percentage of seed dormancy is normal in native forbs.
- The quality of commercially produced native seed has improved dramatically over the past few years.
- Expect your seed to be delivered with certification tags and seed-test results attached.

Designing Seed Mixes

DAVE WILLIAMS

Overview

A well-planned seed mix is essential to reconstructing a diverse and stable plant community. Selecting species for any native planting involves knowing the physical characteristics of the site (soil type, hydrology, slope, aspect, and sunlight exposure), then choosing the most appropriate native plants for that site. All native plantings should include grasses, sedges, and forbs (both legume and nonlegume species). The seed mix should also include annual, biennial, and perennial species to foster early establishment and maintain long-term diversity. Cost and availability of the seed are often the primary factors in determining which species and how much seed of a species to include in a seed mix. Seeding calculators are valuable tools to enable practitioners to develop diverse seed mixes within a seed budget. It is also important to consider the source of seed (see chapter 2) and the ratio of forb to grass species in the mix. A well-planned seed mix will result in a diverse, weed-resistant prairie plant community that will last a lifetime.

Criteria for Species Selection

Selecting the appropriate species of native plants is one of the first steps in planning a reconstruction project. This requires information on site conditions (soil type, moisture, slope, aspect), species characteristics (geographic distribution, light requirements, life span, phenology), appropriate seed sources, and seed cost. All these factors affect the kind and number of species included in a seed mix.

SOIL TYPE AND MOISTURE

Each soil type is a unique blend of sand, silt, clay, and organic matter that affects how the soil drains and retains water. Every plant species has evolved to grow within a certain range of soil moisture conditions. Planting species that are best adapted to the soil moisture of the site will insure their persistence. Include only native grasses, sedges, and forbs in the seed mix that match the soil moisture conditions of the site.

There are five general soil moisture categories: wet (hydric), wet-mesic, mesic (moderate), dry-mesic, and dry (xeric). Wet soils include poorly drained and very poorly drained soils that typically have standing water for part or most of the growing season. Examples of Iowa wet soil series include Palms Muck, Clyde, Coland, Colo, Garwin, Maxfield, and Sawmill. Proceed carefully with sites that have wet soils. These areas may harbor prairie remnants because they were typically too wet to farm (see chapters 8 and 9). Wet-mesic soils include somewhat poorly drained, lighter-colored clay soils. Examples of Iowa wet-mesic soils include Amana, Cooper, Floyd, Kensett, and Muscatine. Mesic soils include well-drained and moderately well-drained loamy soils. Examples of Iowa mesic soil series include Bassett, Dinsdale, Kenyon, Olin, Rockton, and Saude. Dry-mesic soils include somewhat excessively drained glaciofluvial, eolian, and thick loess soils. Examples of Iowa dry-mesic soil series include Dickman, Hamburg, Flagler, and Zenor. Dry soils include excessively drained sandy or gravelly soils and shallow loam soils on steep slopes and ridges. Examples of Iowa dry soil series include Bertram, Chelsea, Dickinson, Finchford, Flagler, and Sparta. To determine the soil type of your site, visit with your local Natural Resources Conservation Service office to obtain a soil map or look online at http://www.nrcs.usda.gov/.

SLOPE AND ASPECT

The site conditions on a slope and the direction it faces (aspect) affect the establishment of native plants. The upper portion of a slope is usually drier than the lower portion; south and west aspects are relatively more dry than the north and east aspects at the same elevation. Thus, there is a sorting of species along the moisture gradient from the top to the bottom of a slope and around it as the aspect changes. Vegetative changes due to slope and aspect can be seen in many native plantings in roadside rights-of-way. These sites often transition from dry to mesic to wet soils in a small area, and the changes in species composition associated with those soils' moistures can be dramatic. If the slope is gradual and the changes in moisture conditions can be easily seen, the extra effort to seed only the species that match the moisture condition of the soil may improve establishment of those species. If the soil moisture gradient isn't as apparent, slopes will need to be shotgun-seeded with all species, requiring inclusion of species in the seed mix that match each moisture condition.

GEOGRAPHIC DISTRIBUTION

Select species that are native to the region of the planting site. The best estimates of species distribution have been determined through observing local remnants and herbarium collections of plants found in native prairies. A region can be defined as the home county and the counties around it. If a species is not present in the region of the planting site, it should be left out of the seed mix. It is possible that a species not currently present in a given region may have occurred there at some time, but without evidence of its presence, we recommend that you not plant it. To obtain a list of tallgrass prairie species (grasses, forbs, and sedges) native to your county, visit the USDA NRCS PLANTS website at http://plants.usda.gov/.

LIGHT REQUIREMENTS

Light availability in nature is a continuum from full-sunlight tallgrass prairies to sun flecks in a closed-canopy forest. Select species that are appropriate to the light conditions of the planting site. If the planting site is adjacent to a woodland and is subjected to reduced sunlight, choose species that are adapted to partial shade. Most seed catalogs group prairie species into three light categories: full, partial, and complete shade.

DIVERSITY

A prairie seed mix that includes species from each plant group (warm- and cool-season grasses, legume and nonlegume forbs, and sedges) will result in a stable, weed-resistant plant community that will attract and sustain wildlife (fig. 3-1). Plant diversity and plant invasion are related: Tilman (1997) found that as plant diversity increases in a prairie, invasion by a new plant species into that plant community was less likely to occur. This is one reason why so many species-diverse prairie remnants have been able to exclude many nonnative weeds. We don't know how many prairie plant species are needed in a prairie reconstruction to exclude perennial weeds. However, there is an increased chance that a species-rich prairie planting will eliminate germinating weed seed by being a better competitor for resources. It may be inexpensive on the front end of the project to plant only a few grass and forb species, but eliminating weeds that have invaded a native planting can be difficult and costly down the road (fig. 3-2).

There is also a link between species diversity and the response to climatic extremes. Tilman and Downing (1994) showed that species-diverse plantings had less reduction in biomass and recovered faster from drought than species-poor plantings. This is important because reduced biomass (food and cover)

Fig. 3-1. A species-diverse prairie reconstruction planted in 2001 at Big Woods Lake in Cedar Falls, Iowa. The seed mix included 79 species of grasses, sedges, and forbs; 61 species were detected in 2007. Photo by Dave Williams.

Fig. 3-2. Invasion of Canada thistle into a species-poor native grass planting in Black Hawk County, Iowa. Photo by David O'Shields.

Fig. 3-3. A forb-rich prairie reconstruction provides nectar sources for insects—food for pheasant chicks. Photo by David O'Shields.

could seriously affect wild and domestic animals dependent on plant growth for survival. Species-diverse plantings can also result in increased diversity of other organisms. Ries et al. (2001) found butterflies preferred native plantings that had grasses and forbs over native plantings with only grasses. Adding native forbs to the seed mix attracts insects and wildlife, such as pheasant chicks, that feed exclusively on insects (fig. 3-3). Species-diverse seed mixes should be strongly considered for all native plantings.

PHENOLOGY

Tallgrass prairie plants exhibit a wide range of growth characteristics. These plants have evolved to take advantage of available resources throughout the growing season. Some grasses and sedges germinate, grow, and flower in spring or fall (cool-season plants), while others germinate in late spring and grow and flower in the summer (warm-season grasses). For a prairie planting to resist nonnative weed invasion, the planting must include native species from both cool- and warm-season grasses, forbs, and sedges. Leaving out any of these groups will expose the planting to weed invasion. In the tallgrass prairie region, it is common to have nonnative cool-season grasses like smooth brome and Kentucky bluegrass invade a prairie planting. We often discover that many of these plantings lack native cool-season grasses and sedges to compete with smooth brome and Kentucky bluegrass.

LIFE SPAN

Native plants can be classified as annuals, biennials, and perennials. Annual plants germinate, flower, and die in one growing season. Biennials germinate and remain vegetative in the first year, then flower and die in the second year. The benefits of having native annuals and biennials in the seed mix are threefold. Annuals and biennials readily germinate and grow large in the first growing season, covering more bare soil and thereby reducing the potential soil erosion. In addition, their rapid growth may reduce weed abundance by competing with weeds for resources in years 1 and 2. Finally, native annuals and biennials flower in the second year, creating a native seed bank to recolonize bare soil created by disturbances that could occur in the future. Most native species are perennials; they flower and survive year after year. Perennials provide long-term diversity and stability. As the planting matures, the perennials predominate. To create a persistent and species-diverse native planting, all seed mixes should contain native annuals, biennials, and perennials.

APPROPRIATE SOURCES

The source of seed for reconstructing prairie is a growing concern to resource managers. Seed derived from multiple remnant sources within the region of the planting site may be better adapted to the climate and soils of a site than seed from distant sources (chapter 2). Preliminary research supports this premise. Williams et al. (2006) found that plants grown from switchgrass seed collected from Iowa remnants had greater growth and survivorship when grown in Iowa than did plants grown from seed of switchgrass cultivars originating in Nebraska and South Dakota. Seed used for a native planting should be derived from prairie remnants within the region of the planting site (see chapter 2).

COST

Another critical factor influencing the number and kind of species included in a seed mix is the cost of the seed. The commercial price for native prairie seed varies greatly by species. In 2007, prairie phlox seed cost $900 per pound compared to gray-headed coneflower seed at $45 per pound. Deciding how many of the more expensive forbs to include depends upon the seed budget and the preference of the person paying for the seed. Consider including some expensive forbs in seed mixes; costs can be controlled by lowering seeding rates of the expensive species. Government programs, such as the United States Department of Agriculture Wetlands Reserve Program, Wildlife Habitat Incentive Program, and Conservation Reserve Program, will have seed costs that vary with program criteria, objectives, and landowner resources. We advocate planting for maximum diversity. It is better to include more species in the seed mix and lower seeding rates than to plant fewer species with higher seeding rates. See tables 3-1, 3-2, and 3-3 for a diverse seed mix for wet-mesic, mesic, and dry sites in Iowa.

Cover Crops and Nurse Crops

A cover crop can be used to seed a site if it is not an appropriate time to seed natives (mid to late summer, for instance). Practitioners often plant nonnative annual cereal crops that can quickly stabilize the soil and reduce weeds until the time is more appropriate to seed natives. For cover crop species, seeding rates, and seeding times, see chapter 12.

Nurse crops or companion crops are usually cereal crops that are planted with the natives and seeded at a lower rate than when planted as a cover crop. The readily germinating seed and quickly maturing plants make nurse crops good competitors against weeds and effective at holding the soil in place on erodible sites while native seedlings are getting established. Due to their life history characteristics, nurse crops tend to diminish from the prairie planting by the third year. Many practitioners already include native nurse crop species in their seed mixes. For nurse crop species, seeding rates, and seeding times, see chapter 12.

Prepackaged Seed Mixes

Many prepackaged native prairie seed mixes are commercially available. Scrutinize each species in the seed mix. Check the label to verify that each species is native to the planting site and that the seed originated from multiple remnant sources from the region of the planting site. Reputable native seed nurseries can provide you with seed-source information if it is not listed on the tag. If you cannot determine the origin of the seed, don't buy it. Avoid mixes that have generic titles such as Midwest wildflowers, wildflower mix, and Midwest adapted.

Developing a Seed Mix Using the Seed Calculator

High-quality seed of hundreds of native prairie species is commercially available today (see the Seed Quality section in chapter 2). The cost of seed varies greatly by species, but forb seed usually adds significantly to the cost of a seed mix. The high cost of native seed has forced us to develop a more precise

Table 3-1. Recommended Native Seed Mix for 1-Acre Wet-Mesic Soil Sites in Iowa (Except Loess Hills)

Species	Seeds/Ounce	Moisture Class	Seeds/ Square Foot	PLS (Ounces)
Grasses/sedges				
Big bluestem	10,000	W-M-D	3.00	13.07
Bluejoint grass	248,880	W-M	3.00	0.53
Broom sedge	84,000	W-M	3.00	1.56
Awlfruit sedge	34,000	W-M	2.00	2.56
Brown fox sedge	100,000	W-M	3.00	1.31
Virginia wild rye	4,200	W-M	1.00	10.37
Fowl manna grass	160,000	W-M	3.00	0.82
Switchgrass	16,000	W-M-D	0.50	1.36
Prairie cord grass	6,040	W-M	2.00	14.42
TOTAL			20.50	45.99
Forbs				
Indigo bush	3,700	W-M	0.10	1.18
Swamp milkweed	4,800	W-M	0.50	4.54
New England aster	67,500	W-M-D	0.50	0.32
Milk vetch	18,662	W-M-D	1.00	2.33
Prairie Indian plantain	4,700	W-M	0.25	2.32
Showy tick trefoil	5,500	M	0.25	1.98
Rattlesnake master	7,500	W-M-D	0.25	1.45
Joe Pye weed	95,000	W-M	1.00	0.46
Boneset	160,000	W	1.00	0.27
Grass-leaved goldenrod	200,000	W-M	2.00	0.44
Bottle gentian	280,000	W-M	2.00	0.31
Wild geranium	5,000	W-M	0.50	4.36
Sneezeweed	130,000	W-M	3.00	1.01
Saw-tooth sunflower	13,440	W-M-D	0.25	0.81
Ox-eye sunflower	6,300	W-M	0.50	3.46
Prairie blazing star	11,000	M	0.25	0.99
Great blue lobelia	500,000	W-M	2.00	0.17
Fringed loosestrife	39,000	W-M	0.20	0.22
Winged loosestrife	9,000	W-M	0.25	1.21
Wild bergamot	75,000	W-M-D	1.00	0.58
Common mountain mint	220,000	W-M	2.00	0.40
Giant goldenrod	250,000	W-M	1.00	0.17
Purple meadow-rue	11,000	W-M	0.50	1.98
Hoary vervain	93,000	W-M	1.00	0.47
Ironweed	24,000	W-M	0.25	0.45
Culver's root	800,000	W-M	2.00	0.11
American vetch	5,000	W-M	0.25	2.18
Golden alexanders	11,000	W-M	1.00	3.96
TOTAL			24.70	36.95

Table 3-2. Recommended Native Seed Mix for 1-Acre Mesic Soil Sites in Iowa (Except Loess Hills)

Species	Seeds/Ounce	Moisture Class	Seeds/ Square Foot	PLS (Ounces)
Grasses/sedges				
Big bluestem	10,000	W-M-D	3.00	13.07
Side-oats grama	8,650	M-D	2.00	10.07
Bebb's sedge	34,000	M	1.50	1.92
Bicknell's sedge	17,000	M	0.50	1.28
Shortbeak sedge	29,000	M-D	0.50	0.75
Troublesome sedge	25,000	M-D	0.50	0.87
Canada wild rye	6,200	M-D	1.00	7.03
Switchgrass	16,000	W-M-D	1.00	2.72
Little bluestem	15,300	M-D	4.00	11.39
Indian grass	11,500	M-D	3.00	11.36
Tall dropseed	30,000	M-D	3.00	4.36
Prairie dropseed	16,000	M-D	0.10	0.27
TOTAL			20.10	65.09
Forbs				
Leadplant	17,884	M-D	0.25	0.61
Canada anemone	8,000	M-D	0.10	0.54
Thimbleweed	16,485	M-D	0.05	0.13
White sage	250,000	M-D	1.00	0.17
Butterfly milkweed	3,350	M-D	0.10	1.30
Milk vetch	18,662	W-M-D	1.00	2.33
White wild indigo	1,700	M-D	0.05	1.28
Partridge pea	2,700	M-D	0.50	8.07
Prairie coreopsis	11,000	M-D	0.05	0.20
Purple prairie clover	18,950	M-D	2.00	4.60
Showy tick trefoil	5,500	M	0.25	1.98
Pale purple coneflower	5,300	M-D	0.75	6.16
Rattlesnake master	6,816	W-M-D	0.50	3.20
Ox-eye sunflower	6,300	M-D	1.00	6.91
Round-headed bush clover	9,000	M-D	0.25	1.21
Prairie blazing star	11,000	M	0.50	1.98
Wild bergamot	75,000	W-M-D	1.00	0.58
Stiff goldenrod	41,000	M-D	1.00	1.06
Prairie phlox	19,000	M-D	0.10	0.23
Common mountain mint	220,000	W-M	1.00	0.20
Gray-headed coneflower	30,000	M-D	1.00	1.45
Black-eyed Susan	84,064	M-D	1.00	0.52
Compass plant	660	M-D	0.10	6.60
Showy goldenrod	103,600	M-D	2.00	0.84

Table 3-2. (*Continued*)

Species	Seeds/Ounce	Moisture Class	Seeds/ Square Foot	PLS (Ounces)
Smooth blue aster	48,000	M-D	2.00	1.82
New England aster	61,628	W-M-D	1.00	0.71
Prairie spiderwort	10,000	W-M-D	0.10	0.44
Hoary vervain	28,000	M-D	0.10	0.16
Culver's root	730,500	W-M	1.00	0.06
Golden alexanders	9,950	W-M	1.00	4.38
TOTAL			20.75	59.71

method of designing seed mixes. The Natural Resources Conservation Service in Iowa has developed a seed calculator program to design seed mixes based upon the number of seeds per square foot for each species. For a free copy of the seed calculator program, visit their website at http://www.ia.nrcs.usda.gov/. The seed calculator will automatically calculate the cost of seeds based upon species selected and their cost. Seed cost can be rapidly recalculated with changes in species selection and seed quantities using the seed calculator. Seeding rate is the total number of live seeds sowed per unit area (seeds per square foot). When using the seed calculator, consider the following recommendations.

All prairie reconstructions should be planted with a minimum of 40 seeds per square foot. Planting fewer than 40 seeds per square foot may result in a weedy plant community. For slopes of 3:1 or greater, we recommend 60 to 80 seeds per square foot because of potential seed loss due to erosion.

Always use a nurse crop on erodible sites.

Develop a species-diverse seed mix. Include a minimum of 6 grasses (cool- and warm-season), 3 sedges, and 25 forbs (5 legume and 20 nonlegume species).

A 50:50 mix of grass seed to forb seed will produce a prairie planting rich in forbs. Therefore, if the seeding rate is 40 seeds per square foot, 20 seeds per square foot are grass and sedge seed, and 20 seeds per square foot are forb seed.

Choose grass, sedge, and forb species native to your region and most appropriate for the soil moisture conditions of the site.

Include annual, biennial, and perennial forb species in the seed mix. Generally, 1 seed per square foot of native annuals and biennials will result in many adult plants. Annuals and biennials should not exceed 10 percent of the total forb seed. Try to equalize the number of seeds per square foot of the perennial forbs as much as your budget will allow.

Table 3-3. Recommended Native Seed Mix for 1-Acre Dry Soil Sites in Iowa (Except Loess Hills)

Species	Seeds/Ounce	Moisture Class	Seeds/ Square Foot	PLS (Ounces)
Grasses/sedges				
Big bluestem	10,000	W-M-D	1.00	4.4
Side-oats grama	8,650	M-D	3.00	15.1
Bebb's sedge	34,000	M	1.50	1.92
Shortbeak sedge	29,000	M-D	0.50	0.8
Heavy sedge	12,000	M-D	0.50	1.8
Troublesome sedge	25,000	M-D	0.50	0.9
Scribner's panic grass	9,000	M-D	0.75	3.6
Canada wild rye	6,200	M-D	1.00	7.0
Porcupine grass	2,132	D	0.10	2.0
June grass	400,000	D	2.00	0.2
Switchgrass	16,000	W-M-D	0.50	1.4
Little bluestem	15,300	M-D	4.00	11.4
Indian grass	11,500	M-D	3.00	11.4
Tall dropseed	30,000	M-D	3.00	4.4
Prairie dropseed	15,000	M-D	0.25	0.7
TOTAL			20.10	65.0
Forbs				
Leadplant	17,884	M-D	0.25	0.6
Canada anemone	8,000	M-D	0.10	0.5
Thimbleweed	16,485	M-D	0.10	0.3
White sage	250,000	M-D	2.00	0.3
Butterfly milkweed	3,350	M-D	0.25	3.3
Whorled milkweed	11,000	M-D	0.10	0.4
Milk vetch	18,662	W-M-D	1.00	2.3
Cream false indigo	1,400	M-D	0.05	1.6
False boneset	32,000	M-D	0.20	0.3
Partridge pea	2,700	M-D	0.25	4.0
New Jersey tea	7,600	M-D	0.20	1.1
Prairie coreopsis	11,000	M-D	0.10	0.4
White prairie clover	18,230	M-D	1.00	2.4
Pale purple coneflower	5,300	M-D	0.25	2.1
Tall boneset	50,000	M-D	0.25	0.2
Flowering spurge	8,000	M-D	0.10	0.5
Prairie sunflower	4,480	M-D	0.25	2.4
Ox-eye sunflower	6,300	M-D	0.50	3.5
Alumroot	700,000	M-D	0.10	0.0
Round-headed bush clover	9,000	M-D	0.50	2.4
Rough blazing star	15,500	M-D	1.00	2.8

Table 3-3. (*Continued*)

Species	Seeds/Ounce	Moisture Class	Seeds/ Square Foot	PLS (Ounces)
Wild bergamot	75,000	W-M-D	2.00	1.2
Stiff goldenrod	41,000	M-D	1.00	1.1
Prairie phlox	19,000	M-D	0.10	0.2
Tall cinquefoil	230,000	M-D	0.10	0.0
Gray-headed coneflower	30,000	M-D	1.00	1.5
Black-eyed Susan	92,000	M-D	1.00	0.5
Wild petunia	5,200	M-D	0.10	0.8
Compass plant	660	M-D	0.10	6.6
Old field goldenrod	300,000	D	1.00	0.1
Showy goldenrod	103,600	M-D	1.00	0.4
Smooth blue aster	52,670	M-D	2.00	1.7
New England aster	67,500	W-M-D	1.00	0.6
Silky aster	476,000	D	1.00	0.1
Prairie spiderwort	8,000	M-D	0.10	0.5
TOTAL			20.05	46.8

Consider including some expensive species that are appropriate for the site at a seeding rate that you can afford. Typically, a species' cost reflects how difficult it is to produce its seed (see chapters 15 and 16). However, cost should not be the only reason for including the species in the seed mix. Therefore, if the site conditions are appropriate, add a little cream false indigo, prairie phlox, or flowering spurge. A small amount of seed is better than no seed.

For a dormant planting, increase the seeds per square foot of warm-season grass species by 25 percent due to increased seed mortality (Henderson and Kern 1999; Meyer and Gaynor 2002). For example, if the seed mix contains 20 grass/sedge seeds per square foot for a nondormant seeding, 25 seeds per square foot should be used for dormant seeding. A possible exception to this may be switchgrass, which has a hard seed coat that can overwinter. The seeding rate for forbs need not be changed for a dormant seeding.

Planting seed at the proper depth (⅛ to ¼ inch) and insuring good seed-to-soil contact are essential for any seed to germinate and establish. Proper seed placement is less certain when broadcast-seeding (hydroseeding, hand-seeding, and broadcast/drop seeders). If broadcast-seeding methods are used, seeding rates for grasses, sedges, and forbs should be increased up to 30 percent (Henderson and Kern 1999).

Summary

- Planting species adapted to the soil moisture conditions of the site will insure their establishment and persistence in the planting.
- Slopes can have variable moisture conditions. Plant species adapted to each condition.
- Select species that are native to the region of the planting site.
- Plant species that are appropriate to the light conditions of the site.
- A diverse seed mix will result in a stable, persistent plant community that will attract and sustain wildlife.
- All seed mixes should contain grasses, sedges, and forbs; include both legume and nonlegume forbs.
- All seed mixes should contain annual, biennial, and perennial forbs.
- Seed of annual and biennial forbs should not exceed 10 percent of the total forb seed.
- Plant seed derived from prairie remnants within the region of the planting site.
- Expensive forbs should not be excluded from seed mixes; costs can be controlled by lowering seeding rates. A small amount of seed is better than no seed.
- Design a seed mix with the goal of creating a prairie planting with maximum species diversity.
- Seed calculator programs can automatically calculate the cost of seed for a given area and recalculate seed cost with changes in species selection and seed quantities.
- Prairie reconstructions should be planted with a minimum of 40 seeds per square foot. For slopes of 3:1 or greater, we recommend 60 to 80 seeds per square foot because of potential loss due to erosion.
- For a diverse planting, we recommend 6 grasses (cool- and warm-season), 3 sedges, and 25 forbs (5 legume and 20 nonlegume species).
- Plant at least a 50:50 ratio of grass seed to forb seed.
- For a dormant planting, increase the seeds per square foot of warm-season grass species by 25 percent (except switchgrass). The seeding rate for forbs need not be changed for a dormant or nondormant seeding.
- Plant seed at the proper depth and insure good seed-to-soil contact. If broadcast-seeding methods are used, seeding rates should be increased up to 30 percent.
- Avoid mixes that have generic titles like Midwest wildflowers, wildflower

mix, and Midwest adapted. Naturalized plant species are nonnative and should not be planted.

- Cover crops are planted to stabilize the soil and reduce weeds when seeding times are inappropriate for seeding natives. For information about cover crops, seeding times, and seeding rates, see chapter 12.
- Always include a nurse crop in the seed mix with the native seed to reduce soil loss on erodible sites. For information about nurse crops, seeding times, and seeding rates, see chapter 12.

Implementing Reconstruction

Site Preparation

DAVE WILLIAMS

Overview

Site preparation involves altering the existing vegetation and soil structure in advance of seeding. The goal is to increase emergence, growth, and survival of the seeded natives by removing thatch, improving seed-to-soil contact, and reducing weeds. From construction sites to cornfields, site conditions can be drastically different and require specific site-preparation techniques. There are three broad categories of site preparation: one associated with bare-soil sites, one associated with sites having vegetation, and a third set that may serve other special needs, such as having a very small planting site and/or a personal preference not to use pesticides. In addition, choose a site-preparation technique that best fits with the type of seeding equipment (see chapter 5) available for the planting project. Choosing an inappropriate technique can result in increased soil erosion, low establishment of the native plants, and an infestation of weeds.

Caution: Most of the site-preparation methods described in this chapter involve killing the existing vegetation and should not be used if remnant prairie vegetation (prairie species that were not planted) is present at the site. The restoration techniques found in chapter 9 should be applied if remnant prairie plants are present.

Removing Trees and Shrubs

The species composition of a planted prairie will change over time if volunteer trees and shrubs are not eliminated. Shade from trees and shrubs will create cooler and moister conditions under the canopy, favoring shade-tolerant plant species and displacing full-sunlight prairie species. In addition, woody plants that are left on the site will spread, by suckering and seed, further displacing

Table 4-1. Persistent Perennial Plants That Should Be Killed Prior to Planting Prairie Vegetation

Species	Phenology	Herbicide Class	Application Method	Application Time
Kentucky bluegrass	grass	glyphosate	foliar applied	in spring at boot-to-early seed-head stage
Quackgrass	grass	glyphosate	foliar applied	6–8" tall
Reed canary grass	grass	glyphosate	foliar applied	in spring at boot-to-early seed-head stage
Smooth brome	grass	glyphosate	foliar applied	in spring at boot-to-early seed-head stage
Tall fescue	grass	glyphosate	foliar applied	in fall with 6–12" new growth
Bird's-foot trefoil	herbaceous	clopyralid, triclopyr	foliar applied	up to 5 leaf
Canada thistle	herbaceous	clopyralid	foliar applied	in spring, rosette to bud
Crown vetch	herbaceous	triclopyr	foliar applied	up to 5 leaf
Leafy spurge	herbaceous	picloram	foliar applied	in spring at true flower stage or fall regrowth
Black locust	woody	triclopyr	cut stump or basal bark	anytime except with snow or running water
Boxelder	woody	picloram, triclopyr	cut stump or basal bark	anytime except with snow or running water
Common buckthorn	woody	picloram, triclopyr	cut stump or basal bark	anytime except with snow or running water
Gray dogwood	woody	picloram, triclopyr	cut stump or basal bark	anytime except with snow or running water
Green ash	woody	picloram, triclopyr	cut stump or basal bark	anytime except with snow or running water

Table 4-1. (*Continued*)

Species	Phenology	Herbicide Class	Application Method	Application Time
Honey locust	woody	triclopyr	cut stump or basal bark	anytime except with snow or running water
Multiflora rose	woody	picloram, 2, 4-D	cut stump or basal bark	anytime except with snow or running water
Eastern red cedar	woody	none needed		
Siberian elm	woody	picloram, triclopyr	cut stump or basal bark	anytime except with snow or running water
Silver maple	woody	picloram, triclopyr	cut stump or basal bark	anytime except with snow or running water
Smooth sumac	woody	triclopyr	cut stump or basal bark	anytime except with snow or running water
Tartarian honeysuckle	woody	picloram, triclopyr	cut stump or basal bark	anytime except with snow or running water

Note: Always read and follow label directions.

prairie plants. Woody plants can also interfere with no-till drilling and broadcast seeding of the natives.

Some practitioners leave a few clone groups of native trees like wild plum and choke cherry on the site for habitat. However, native trees and shrubs like boxelder, red cedar, and gray dogwood should be removed because they can aggressively spread in a planted prairie (table 4-1). We recommend the removal of all nonnative trees and shrubs.

Herbicides are very effective at killing woody plants. Smaller trees and shrubs can be foliar-sprayed. Trees greater than half an inch in diameter need to be cut and the stump chemically treated to prevent resprouting (table 4-2). Herbicides should be applied only to the inner bark (cambium layer) of the cut surface (see fig. 6-3). The inner bark region of the cut stump is a thin layer adjacent to the outer bark of the tree. Because of the high concentration of chemicals in stump-treatment herbicides, it is important to be careful not to dribble herbi-

Table 4-2. Recommended Herbicides for Control of Trees and Shrubs

Herbicide Class	Trade Name	Residue Activity	Cut Stump	Foliar	Apply Near Water
Fosamine ammonium	Krenite S	none	no	yes	no
Glyphosate	Rodeo	none	yes	yes	yes
Glyphosate	Roundup	none	yes	yes	no
Picloram + 2,4-D	Pathway	noncrop sites only	yes	no	no
Triclopyr	Element 3A	3 weeks	yes	yes	yes
Triclopyr	Garlon 4	noncrop sites only	yes	yes	no
Triclopyr	Pathfinder II	noncrop sites only	yes	no	no
Triclopyr + 2,4-D	Crossbow	3 weeks	yes	yes	no

Note: Always read and follow label directions.

cides off the cut surface onto the ground. Coniferous trees (pines and cedars) do not need to be treated after being cut because they will not resprout, but all deciduous trees need to be treated. Stumps need to be cut flat and as close to the ground as possible to prevent interference with seeding equipment.

Herbicides and Their Residue

Herbicide residue can inhibit establishment of native grasses and forbs up to 4 years after application (table 4-3). When planting natives, it is important to know what kinds of herbicides have been previously used on the site. Review pesticide records to learn what herbicides have been used on a particular site. Pesticide application records identify the herbicide used, when it was applied, and the application rate. On sites previously used for agricultural crops, contact the farmer to obtain pesticide application records. Most states require pesticide applicators to retain pesticide records for several years.

Herbicide labels have information concerning which crops can be safely planted after spraying. Unfortunately, native grass and wildflower species are not included on most herbicide labels. Do not assume that a native species in the same genus as a labeled species will tolerate the herbicide. A conservative approach is to plant natives no sooner than the time listed for the most sensitive crop on the label. Most herbicide labels can be found on the Crop Data Management Systems website, www.cdms.net.

Table 4-3. Length of Time That Herbicides Can Remain Active in the Soil after Application

Herbicide Class	Trade Name	Residue Activity (for Labeled Crops)	Weeds Controlled
2,4-D	Weedone 650	3 months	annual/perennial broadleaves
Acetochlor	Degree	rotation only to soybeans, corn, milo, wheat, tobacco	annual grasses and annual broadleaves
Alachlor	Micro-Tech	rotation to crops not listed on the label is prohibited	annual grasses and annual broadleaves
Atrazene	Aatrex 4L	1 year	annual grasses and annual broadleaves
Bentazon	Basagran	none	annual/perennial broadleaves
Bromoxynil	Buctril	1 month (corn, sorghum, small grains)	annual/perennial broadleaves
Chlorsulfuron	Telar DF	6 months (timothy)	annual/perennial broadleaves
Clopyralid	Stinger	18 months (peas, potatoes)	annual/perennial broadleaves
Clopyralid	Transline	conduct field bioassay for symptoms on selected crop	annual/perennial broadleaves
Dicamba	Banvel	3 years	annual/perennial broadleaves and woodies
Fluazifop	Fusilade DX	2 months (corn, sorghum, cereal crops)	annual/perennial grasses
Fosamine ammonium	Krenite S	none	woodies
Glyphosate	Roundup	none	all grasses, broadleaves, and woodies
Imazamox	Raptor	3–26 months (various crops listed)	annual grasses/ broadleaves, some perennials

Table 4-3. (*Continued*)

Herbicide Class	Trade Name	Residue Activity (for Labeled Crops)	Weeds Controlled
Imazapic	Plateau	12–48 months depending upon application rate	annual/perennial grasses
Imazethapyr	Pursuit	40 months	annual grasses and annual broadleaves
Metolachlor	Dual II Magnum	9 months (clover)	annual grasses and annual broadleaves
Oryzaline	Surflan AS	1 year	annual grasses and annual broadleaves
Pendimethalin	Pendulum 2G	3 months (turf grasses)	annual grasses and annual broadleaves
Pendimethalin	Prowl 3.3 EC	12 months (beets, spinach)	annual grasses and annual broadleaves
Picloram	Tordon 22K	2 years for small grains	perennial broadleaves and woodies
Picloram + 2,4-D	Pathway	not for use on cropped sites	woodies
Sethoxydim	Poast	4 months	annual/perennial grasses
Simazine	Simazine 4L	1 year (grasses, legumes)	annual grasses and annual broadleaves
Sulfometuron	Oust XP	6 months (smooth brome)	annual/perennial grasses and broadleaves
Triclopyr	Element 3A	3 weeks	annual/perennial broadleaves and woodies
Triclopyr	Garlon 4	not for use on cropped sites	annual/perennial broadleaves and woodies
Triclopyr	Pathfinder II	not for use on cropped sites	woodies
Triclopyr + 2,4-D	Crossbow	3 weeks	annual/perennial broadleaves and woodies
Trifluralin	Treflan	12 months spring applied, 14 months fall applied	annual weeds (grasses/ broadleaves)

Note: Pesticide residue information was compiled from the herbicide trade labels listed in the table. Residue activity time is after application of the herbicide. Activity can vary depending on the rate, soil, climate, and sensitivity of the species.

Site Preparation for Bare-Soil Sites

Three kinds of bare-soil sites are commonly found throughout the tallgrass prairie region. In this section, we describe the best site-preparation treatments for each. Some of these options will depend upon the availability of specialized seeding equipment (see chapter 5). A caution statement is included with those options that should not be used on erosive sites.

CONSTRUCTION SITES

In many construction sites, the original soil profile has been altered during the construction process. Typically, topsoil is removed during construction and replaced when the project is finished. Any vegetation present on the site is likely to be annual weeds germinated from the seed bank in the replaced soil. Some areas within the site may have compacted soil from construction equipment.

Construction sites may require site preparation at any time during the year. This adds an additional level of complexity to the reconstruction project. The optimum times to seed natives (see chapter 8) may not coincide with times when the construction project is ready for site preparation. The practitioner must decide what to do with the site in the meantime. A sense of urgency is added if the site is susceptible to soil erosion.

An ideal seedbed for a native seeding should consist of friable soil particles (half an inch or smaller) in the top inch of soil. Many construction sites have large clods and compacted soil. To assess soil compaction, use the screwdriver method described later in this chapter. If the soil surface is not compacted and no large clods (greater than half an inch) are present, the site is ready for seeding. Use of a no-till seed drill requires a firm seedbed before seeding. If a footprint impression can be made half an inch or deeper in the soil, the site needs to be cultipacked (rolled) before seeding with a no-till drill (fig. 4-1). However, if a drop seeder or a broadcast seeder is used for seeding, the site should be cultipacked only after seeding. We recommend seeding with a no-till seed drill, but any seeding method can be used (see chapter 5).

ROW-CROPPED SITES

Sites that have been row-cropped may require different kinds of site preparation. After harvest, some fields are not tilled and crop stubble and residue remain, while other fields may have been tilled and little crop residue remains on the soil surface. Crop residue can range from fine stubble (soybeans) to very coarse stubble and dense crop residue (corn). The amount of site preparation needed depends upon the quantity of crop residue left on the field and the type

Fig. 4-1. Cultipacking can break apart large clods and improve the seedbed prior to seeding. Cultipacking after seeding can improve seed-to-soil contact. Photo by Dave Williams.

of seeding equipment used for the project (see chapter 5). See the section on assessing crop residue later in this chapter.

Option 1, light crop residue and no-till drill seeding: This method is a rapid process for site preparation and may reduce soil erosion because no tillage is needed before seeding. This option is recommended if the site has light crop residue and is seeded with a no-till seed drill. No other site preparation is needed. Native seed can be no-till-drilled directly into the crop residue.

Option 2, light crop residue and drop or broadcast seeding: This method can be very economical if access to a no-till grass drill is not possible. This option is recommended if the site has light crop residue and is seeded with a drop seeder or a broadcast seeder. The site should be cultivated with a spike-tooth harrow before seeding. One pass with the harrow should break apart the crop residue that is matted on the surface and loosen the soil at the surface for improved seed-to-soil contact. This option requires cultipacking the site after seeding. Cultipacking presses the seed into the soil, improving seed-to-soil contact.

Option 3, heavy crop residue and any seeding method: The benefit of this method is that the soil will be conditioned to maximize seed-to-soil contact and may improve native plant emergence (Morgan 1997). This option is recommended if the site has heavy crop residue and is seeded with a no-till drill, drop seeder, or broadcast seeder. Caution: Do not remove heavy residue by prescribed burning, because the fire can be difficult to control and extinguish. First, mow (chop) the stalks. Then, disk the site until most of the crop residue disappears (figs. 4-2 to 4-5). Mowing the stalks prior to disking should reduce the disking effort. This site-preparation technique requires cultipacking. When to cultipack the site depends upon the type of seeding equipment used. Cultipack the site before seeding if a no-till seed drill is used. Cultipack the site after seeding if the site is seeded with a drop seeder or a broadcast seeder. Caution: This option is not recommended for erosive sites. A seed drill requires a firm seedbed before seeding. We recommend seeding with a no-till seed drill, but any seeding method can be used (see chapter 5).

FEEDLOTS AND OVERGRAZED PASTURES

Bare soil can result from severe overgrazing and trampling by livestock. However, this bare-soil condition may be temporary. Removing livestock from the site can cause vegetation to reemerge from underground rootstock. We recommend that livestock be removed for at least one entire growing season to allow the vegetation to recover and be identified. If remnant (not planted) prairie plants are detected, the site should be considered a prairie remnant, and site-

Fig. 4-2. Disking a cornfield. Corn stover can be reduced from the planting site by disking. At least three disking passes are needed to adequately condition the soil prior to seeding the natives. Photo by David O'Shields.

Fig. 4-3. One disking pass. Large clods and corn stover remain on the surface after the first disking pass. Photo by David O'Shields.

Fig. 4-4. Second disking pass. The second disking pass greatly reduces the amount of corn stover and the size of clods. Photo by David O'Shields.

Fig. 4-5. Third disking pass. After the third disking pass, most of the corn stover is tilled into the soil, and the clods are broken into small pieces. The site is ready for seeding natives. Photo by David O'Shields.

preparation techniques for prairie remnants should be used (see chapter 9). Typically, however, feedlots and overgrazed pastures will contain persistent perennial plants and high levels of weed seed in the soil. Manure can also contribute to high levels of nitrogen in the soil, which will stimulate weed germination and weed growth. For this reason, these sites require a unique site-preparation and management strategy focusing first on weed control (see table 4-1). For sites like these, with an abundance of weeds, we recommend site-preparation techniques from the Stand Replacement subsection of this chapter.

Site Preparation for Vegetated Sites

Selecting a site-preparation technique for a vegetated site depends upon the kinds of plants established, whether the existing vegetation was planted or is remnant, and the goal of the planting project. Types of vegetated sites include turf grass lawns, pastures, hayfields, and conservation plantings. Current vegetation on these sites can vary from a stand consisting of smooth brome/alfalfa hayfields and Kentucky bluegrass lawns to a dense stand of prairie grasses on a site enrolled in a federal Conservation Reserve Program. Caution: If a site contains remnant (not planted) prairie plants, the site should be considered a prairie remnant, and site-preparation techniques for prairie remnants should be used (see chapter 9).

There are two groups of site-preparation options: stand replacement (starting over) and stand enhancement (interseeding). Stand replacement site-preparation techniques should be used when the goal is to replace a current stand of nonnative grasses and legumes with prairie grasses and wildflowers. Stand enhancement site-preparation techniques are typically used when the goal is to add prairie species to sites that currently have some native plants or are dense stands of prairie and pasture grasses with few to no wildflowers.

STAND REPLACEMENT (STARTING OVER)

The goal of stand replacement is to convert an established stand of nonnative plants into a native stand of grasses and forbs. Stand replacement has three primary methods of site preparation. Select a site-preparation technique based upon the speed with which you want to complete the project, the budget for the project, and the kind of equipment available to conduct site-preparation activities.

Option 1, spray and plant: This method is a rapid process for site preparation which may not eliminate all the existing vegetation but is very economical. Site preparation begins by reducing standing dead material and thatch by mowing or burning. Mow (4 inches high or less) in spring or late summer, or burn when the vegetation is dormant. Apply an appropriate herbicide to actively growing vegetation when there is 4 to 6 inches of new growth (see table 4-1, fig. 4-6). For legume/grass stands, a mixture of a broadleaf and grass herbicide such as glyphosate and 2,4-D should be used. It can take 2 to 4 weeks after mowing or burning for the vegetation to have enough new growth for herbicide treatment. Respray any green plants after 14 days from the first herbicide treatment. Wait another 14 days after the last herbicide treatment to seed. Seed can then be broadcast or drilled (see chapter 5).

Option 2, repeated spray and plant: This site-preparation technique requires an entire growing season and is more expensive than option 1, but control of persistent perennial plants is greatly improved. Site preparation begins by reducing standing dead material and thatch by mowing or burning. Mow (4 inches high or less) or prescribe-burn in early spring. Apply glyphosate to actively growing vegetation when there is 4 to 6 inches of new growth; respray or spot-treat each time it greens up with 4 to 6 inches of new growth throughout the summer and into early fall. The site will be ready for a dormant or nondormant seeding the following year (see chapter 5). No further site preparation is needed if the site is seeded with a no-till seed drill. However, if the site is to be

Fig. 4-6. A small (25-gallon) electric boom sprayer can be used to spray 1- to 2-acre sites. Photo by David O'Shields.

seeded with a broadcast seeder or a drop seeder, the area should be roughed up with a spike-tooth harrow before seeding. This will break apart any thatch and loosen the soil at the surface for improved seed-to-soil contact. After seeding, the site should be cultipacked.

Option 3, spray, till, and plant: This method controls both the established persistent perennial plants and germinating weed seed in the soil. Caution: This option is not recommended for erosive sites, as repeated disking will create bare soil. Site preparation begins by reducing standing dead material and thatch by mowing or burning. Mow (4 inches high or less) or prescribe-burn in early spring. Apply glyphosate to actively growing vegetation when there is 4 to 6 inches of new growth. Wait 10 days after the herbicide application, and disk the site at 3-week intervals for the entire growing season. After the final disking operation, the next step in conditioning the soil depends upon the seeding method. Use of a no-till seed drill requires a firm seedbed before seeding. If a footprint impression can be made half an inch or deeper in the soil, the site needs to be cultipacked before seeding with a no-till grass drill. However, if a drop seeder or a broadcast seeder is used for seeding, the site should be cultipacked only after seeding. The site can be fall-seeded or seeded in the spring of the following year. We recommend seeding with a no-till grass drill, but any seeding method can be used (see chapter 5).

STAND ENHANCEMENT (INTERSEEDING)

Stand enhancement techniques are often applied to sites that are dominated by grasses with few to no prairie forbs. These plantings can range from dense stands of prairie grasses to nonnative cool-season grasses and legumes. Many of these sites are conservation plantings enrolled in government programs where haying and grazing have been restricted. The goal of stand enhancement is to add native grasses and forbs into an existing planting without eliminating the established vegetation.

Option 1, direct interseeding: This method is least disruptive to the existing plant community and does not require any manipulation of the existing plant community. Native plant germination and establishment can be slow compared to the other options. Seed is sowed into the established vegetation without any manipulation to the established vegetation. Prairie plant establishment can be low in stands of persistent perennial plants, and the nonnative vegetation can persist for many years after seeding the natives. Any seeding method can be used (see chapter 5). This site-preparation option is quick and can be done without any specialized equipment, but it will require patience.

Option 2, repeated mowing and interseeding: This method reduces the competition from existing vegetation and enhances emergence and survival of seeded prairie grasses and forbs by suppressing the existing vegetation with repeated mowing (Williams et al. 2007). Standing dead material and thatch should be removed in the fall either by prescribed burning or by late summer haying. To improve germination of forb seed, a dormant seeding is recommended. Any seeding method can be used; however, we recommend that a no-till grass drill be used for best seed-to-soil contact (see chapter 5). Dormant seeding will permit stratification of the forb seed and improve germination (see chapter 16). The following growing season, from late April to early September, the site should be mowed to the height of 4 inches every two weeks. This frequency of mowing during the spring and summer months removes the dense canopy of cool- and warm-season grasses, allowing sunlight to reach the new seedlings. Mowing also prunes the grass roots by pruning the shoots, freeing up root space for new forb seedlings in the dense grass sod.

Option 3, spray, mow, and interseed: This site-preparation method reduces competition between the seeded grasses and forbs and the existing vegetation by suppressing a portion of the existing vegetation with herbicides (Cohen 2007). Standing dead material and thatch should be removed in the spring by prescribed burning, haying, or mowing. Spray 50 percent of the stand with a grass herbicide (sethoxydim) when there is 4 to 6 inches of new growth. The entire site can be fall-seeded or seeded in early spring of the following year. Any seeding method can be used; however, we recommend that a no-till grass drill be used for best seed-to-soil contact (see chapter 5). Dormant seeding will permit stratification of the forb seed and improve germination (see chapter 16). The following growing season, the entire site should be mowed once in early summer.

Option 4, disk and interseed: This site-preparation method reduces competition between the seeded grasses and forbs and the existing vegetation by suppressing a portion of the existing vegetation with tillage. Standing dead material and thatch should be removed in late summer by grazing, haying, or mowing. Lightly disk (to the 4-inch depth) 50 percent of the site in early fall (USDA 2002). The entire site can be seeded in late fall to early spring. Any seeding method can be used; however, we recommend that a no-till grass drill be used for best seed-to-soil contact (see chapter 5). Dormant seeding will permit stratification of the forb seed and improve germination (see chapter 16). The site can be fall-seeded or seeded in early spring of the following year. Mow the site in late spring and in midsummer of the following year. Additional mowing may be needed to control annual weeds.

Alternative Site-Preparation Methods

Site preparation doesn't have to involve expensive equipment and herbicides. Sod removal and solarization are very effective techniques for removing established nonnative plants and seeds stored in the soil. These techniques may be best suited for smaller sites such as residential yards and areas where equipment access is limited. They may also be useful on projects where chemical treatments are not desired.

SOD REMOVAL

Sod removal is very effective in permanently removing turf grass and weed seed embedded in the sod. Using a sod cutter, cut turf grass sod and remove it from the site. Adjust the machine to cut the sod at the maximum depth in order to remove as much root material and weed seed as possible. Don't discard the cut sod—it may have value! You may be able to pay for the machine rental by selling the cut sod to local contractors or landscaping companies. After a few weeks, when the soil has loosened, prairie seed can be lightly raked into the soil and packed with a lawn roller. Sod-cutting machines are available at equipment rental stores.

SOLARIZATION

Begin by covering the ground with black plastic for 8 weeks during the growing season to kill existing vegetation. The plastic needs to be stapled or weighted to the ground to prevent the wind from blowing it off site. Staples can be made by bending 12-inch sections of number 9 fencing wire around a 1-inch pipe. Use a mallet or a small 2-by-4 piece of wood to push the staples into the ground. Heat generated under the covering will also kill weed seed in the soil. Once the covering is removed, prairie seed can be lightly raked into the soil and packed with a lawn roller. Solarization is a good and popular alternative to using pesticides and tillage to prepare a site for a prairie planting.

The Screwdriver Method for Compacted Soil

Soil compaction develops when there is a decrease in air space between soil particles. Heavy machinery operation and trampling by humans or livestock can cause soil compaction. Compacted soil can severely reduce the establishment of natives by preventing seed from being planted at a proper depth and by inhibiting root penetration of newly germinated seedlings. To check for soil compaction near the soil surface, stick a large, flat screwdriver into the soil at multiple spots in the compacted region. If the screwdriver cannot be pushed more than 2 inches into the soil in at least half of the spots, there is a good chance that the soil surface is too compacted. To eliminate surface compaction, rototill the site to loosen the upper 4 inches of soil. Any large (greater than half an inch) dirt clods need to be broken into smaller pieces. To reduce clod size, harrow the site using a drag harrow or a piece of chain link fence with some weight added.

Assessing Crop Residue

Crop residue can be grouped into two categories: light or heavy. Light crop residue is defined as crop stubble no more than 4 inches high, with residue on the surface not intertwined and some bare soil that can be seen through the residue. Light crop residues can include soybean residue, corn residue after a silage harvest, corn residue after baling the residue, or cereal grain residue. Heavy crop residue is defined as crop stubble taller than 6 inches, intertwined and layered on the surface, with no bare soil visible. Heavy crop residues can include corn or sorghum residue after a typical harvest and some cereal grains, such as winter wheat.

Assessing crop residue in the field must be done onsite. Walk in a line across the field, and stop in ten spots of equal distance from end to end. Look down near your feet to see if bare soil is visible. Reach down and grab some residue; it is intertwined if a layer of residue larger than your hand comes off the ground. If intertwined residue is found and bare soil cannot be seen in more than five spots, consider the crop residue to be heavy.

Summary

- Killing the established nonnative vegetation and conditioning the soil for optimum seeding are the two goals of site preparation.
- To kill persistent perennial plants, multiple herbicide applications and/or multiple tillage operations are needed.
- Site preparation will vary depending upon current land use and the seeding method.
- Use no-till site-preparation options on sites susceptible to soil erosion to minimize soil disturbance.
- If the soil is tilled and a no-till seed drill is used to plant the natives, the site should be cultipacked (rolled) before seeding.
- If the site is tilled and a broadcast seeder or a drop seeder is used to plant the natives, the site should be cultipacked after seeding.
- Removing standing dead material and thatch is critical for native seed to have good seed-to-soil contact.
- Removing sod from the site with a sod cutter will reduce the weed seed bank and reduce weed competition.
- Covering the ground with black plastic for 8 weeks during the growing season will kill the established vegetation and sterilize weed seed in the soil.
- Woody plants can interfere with no-till drilling and broadcast seeding of the natives. Cut trees flat and close to the ground to allow the no-till drill to seed over the stumps.
- To eliminate woody competition with prairie plants, foliar-spray smaller trees and shrubs and cut and stump-treat larger trees.
- If the site was previously used for agricultural crops, check pesticide records for herbicide residue before planting natives.
- Compacted soil can inhibit proper seeding depth and root penetration of seedlings. Tillage may be needed to loosen compacted soil.

Seeding

DAVE WILLIAMS

Overview

Deciding when to plant a prairie is a challenge. Some species establish better when planted in the spring, other species establish better when planted in the fall, and some species are hard to establish whenever they are planted. Seeding rates of some species may need to be increased depending on when and how they are planted.

Prairie seed can be planted by broadcast seeding, hydroseeding, and drill seeding. Regardless of the seeding method used, it is essential that seed be planted at the proper depth and with good seed-to-soil contact. It's the responsibility of the person actually doing the seeding to ensure that seed is planted correctly. This chapter will offer insights into the process of seeding native plants.

Seeding Time

Tallgrass prairie plants exhibit a wide range of growth characteristics. With adequate soil moisture, cool-season grasses and many forbs germinate in early spring when minimum soil temperatures are between 39° to 45° F, while warm-season grasses germinate in late spring when soil temperatures reach 50° to 56° F (Smith et al. 1998). Real-time soil temperatures for the tallgrass prairie region can be found at http://www.greencastonline.com/SoilTempMaps.aspx. Native seed mixes often contain both cool- and warm-season species, and there is no single best time to plant. However, choosing a planting time to maximize germination and establishment depends upon the species selected and their contribution to the seed mix. A seed mix with a strong forb component (50 percent or greater forb seed) should be dormant-seeded. By contrast, a seed mix of mostly warm-season grasses (70 percent or greater grass seed) should be seeded in mid spring.

DORMANT SEEDING (SOIL TEMPERATURES BETWEEN 32° AND 38° F)

A dormant seeding is defined as planting seed during a time when there is the least chance of germination and, thus, the seed will lie dormant for several months. For most of the tallgrass prairie region, dormant seeding can begin in early November. Early onset of very cold weather in the fall or cold weather lasting into late winter can extend the calendar times for dormant seeding. The benefits of dormant seeding are twofold. First, seeding when the soil temperature is below 39° F ensures that there is no germination of the natives until the following spring, when environmental conditions are suitable for germination and growth. Second, dormant seeding benefits forbs by permitting stratification, which improves germination. We recommend that dormant seeding be done only if the seed can be planted into the soil (⅛ to ¼ inch deep) and packed. Broadcasting seed onto ice or frozen ground is not recommended, as this will expose the seed to wind erosion and predation. Dormant planting mimics the natural process of seed ripening and autumn/winter dispersal of many prairie species. However, dormant seeding of most native grasses (except switchgrass and Canada wild rye) increases seed mortality (Meyer and Gaynor 2002). If the seed mix contains 50:50 forb seed to grass seed or greater, dormant seeding should be considered. Grass seed should be increased by 25 percent if dormant-seeded to compensate for seed loss (Henderson and Kern 1999).

SPRING SEEDING (LATE MARCH TO MID JUNE)

There is a wide range of soil temperatures in spring. Spring soil temperatures (1 inch deep) in Iowa range from 35° F in late March to over 70° F in mid June (Riley 1957). The specific time of year that the site is seeded will determine which species are favored in the seed mix. Early spring seeding favors cool-season grasses, sedges, and some forbs. The window for germination of cool-season plants diminishes as soil temperatures increase throughout spring. A late spring seeding favors warm-season grasses and some forbs. Spring seeding may not permit adequate stratification for some forbs to break dormancy. Nongerminated seed will remain in the soil until conditions are appropriate for germination. Some practitioners believe that this nongerminated seed will perish before another opportunity to germinate comes around. To alleviate this problem, artificial stratification of the forb seed has been used by some to increase germination. To artificially stratify seed, refrigerate it in moist sand for 4 to 8 weeks. Use a box fan to dry the seed on a tarp for a few hours just prior to seeding. Plant the seed immediately after drying it. There are some risks with direct-seeding artificially stratified seed. Seed mortality will result if the seed is not planted at the proper depth (⅛ to ¼ inch deep) and/or packed after seeding or if there is not adequate soil moisture at the time of sowing.

SUMMER SEEDING (JULY TO SEPTEMBER)

Planting in mid and late summer is risky business. New germinates exposed to excessive heat and drought will perish. In addition, many prairie species require 2 to 6 weeks to germinate. By the end of the growing season, it is likely that late-emerging seedlings may be too small to survive the winter. Seeding natives during this time is not recommended.

FROST SEEDING (FEBRUARY TO MARCH)

Frost seeding is a special form of dormant seeding done at the tail end of winter, when temperatures are below freezing at night and above freezing during the daytime. If the soil surface is free of snow or ice, seed can either be drilled or broadcast. The freeze-thaw action creates small cracks in the soil and allows seeds to settle into it. The effect on germination of prairie grass and forb seed by a frost seeding compared to other seeding times is unknown. However, research on nonnative legumes has shown that frost seeding can improve seed germination, but an unusually dry and warm spring can result in poor establishment (Barnhart 2002). In addition, the effect on germination of nonnative cool-season grasses that are frost-seeded can vary, and it is not recommended for some species (West et al. 1997). The benefit of frost-seeding prairie seed may be related to the length of time that the seed remains in the soil before germinating. Compared to a dormant seeding in November, frost seeding reduces the time that seed remains in the soil before germination and may reduce seed mortality from pathogens and/or predation (Hemsath 2007). We believe that frost seeding can be a good time to seed for most native seed mixes. We recommend seeding with a no-till grass drill to maximize seed-to-soil contact. If broadcast seeding is used, the seeding rate should be increased by 25 percent to compensate for seed loss due to wind erosion and predation (Henderson and Kern 1999). Frost seeding is not recommended on eroded sites with rills and gullies. If the site is prone to erosion, a temporary cover crop and/or a mulch should be applied to keep the seed in place (see chapter 12).

Seeding Methods

Planting seed at the proper depth with good seed-to-soil contact is essential. Seed planted too deep will not emerge, resulting in poor stand establishment. Likewise, seed not covered by soil can germinate, desiccate, and die. Prairie seed can be properly planted by each seeding technique recommended in this section. However, it's the responsibility of the person actually doing the seeding to insure that seed is planted correctly. This requires periodic checking of the planted seed and the equipment during seeding.

BROADCAST SEEDING

Broadcast seeders range from tractor- and ATV-mounted implements to hand-held seeders or simple hand-broadcast seeding. This method can be a low-cost way to seed your prairie. An inexpensive hand-held fertilizer spreader, available at your local hardware store, can be used for seeding. Mosaic seeding can be easily done by using any broadcast-seeding method (see the Seeding Design section in this chapter).

To insure that the seed is evenly distributed and dispersed over the planting site, the seed must be properly mixed and the seeding rate carefully calculated. The seed should be mixed in equal parts with inert material such as vermiculite, cracked corn, or kitty litter. This will increase the volume of the seed. Because of improvements in seed cleaning, the volume of prairie seed needed to plant a smaller site (1 acre or less) may not fill a 5-gallon bucket. Mixing any of these materials with the prairie seed will improve the seed flow through the seeder and will make calculating the seeding rate much easier. Seed can be mixed in a plastic tub by hand or on a concrete slab using a flat shovel. After thoroughly mixing the seed with the inert material, weigh the mixed seed.

The next step is to calculate the seeding rate. The first step is to determine the seeding width of the seeding equipment. This can be done by running the seeder on a concrete slab with a small amount of seed mix. To determine the seeding width, measure the width that includes most of the seed on the floor. For handheld broadcast-seeding equipment, follow the same step as above. Then calculate how many passes over the site you need for complete seed coverage. Seeding passes can be calculated by dividing the width of the planting area by the seeding width. For example, if the planting site is 100 feet wide and 200 feet in length and the seeding width is 8 feet, divide 100 feet (site width) by 8 feet (seeding width). This suggests that 12.5 seeding passes are needed to cover the entire site. A seeding pass is equal to the length of the planting site. In the example, one seeding pass equals 200 feet.

The next step is to conduct a test seeding on the planting site. Set the feed gate on the seeder to the middle setting if a mechanical seeder is used. Otherwise, grab an equal portion of seed each time you pull it from the container you are carrying. Seed a 20-foot strip on the planting site using a preweighed small quantity of mixed seed. Use a ground speed that is recommended by the seeding equipment manufacturer, or walk a steady pace if hand-broadcasting. Reweigh the remaining seed and subtract it from the preweighed quantity. This will give you the amount of seed (by weight) used for a 20-foot seeding strip. Assume that we used 1 ounce of seed mix for the 20-foot seeding strip.

Next, divide the length of one seeding pass (the length of the planting site) by the 20-foot seeding strip. In this example, 200 feet (site length) is divided by 20 (test seeding strip), which results in 10. Multiply 10 by the quantity used in the test seeding, which was 1 ounce. This will give the amount of seed that you would potentially use when seeding one pass. In our example, we would seed 10 ounces of seed mix for every seeding pass (200 feet). To calculate the total amount of seed mix needed for the entire site, multiply the amount of seed used in one seeding pass by the total number of seeding passes needed to cover the site. In our example, 10 ounces is multiplied by 12.5 seeding passes, which equals 125 ounces of seed mix needed to seed the entire site. If this amount is too high or too low compared to the total amount of seed mix, you can adjust your ground speed and/or the feed gate on your seeder. Reducing your ground speed by half will double the amount of seed mix needed to seed the site. Likewise, doubling your ground speed will reduce the amount of seed mix needed to seed the entire site by half. We recommend using a constant ground speed. If a change needs to be made in the seeding rate, readjust the feed gate on the seeder or the amount you hand-broadcast as needed.

Remember, each time the seeder gate is readjusted, the seeding rate should be recalculated by going through the process of a test seeding again. If the site is hand-broadcast, it may be easier to adjust the seeding rate by adding more inert material to the seed mix to increase its volume as needed. We recommend lowering the seeding rate so the entire site can be covered twice. This will insure even seed dispersal and distribution over the site. After seeding, seed should be incorporated into the soil to improve seed-to-soil contact. Incorporating seed into the soil can be done by dragging a piece of heavy chain or a piece of chain link fence, using a drag harrow, or raking seed in with a garden rake (fig. 5-1). Drag, harrow, or rake until the seed disappears. Finally, pack the soil with a cultipacker or lawn roller (fig. 5-2).

GRASS DRILLS

A grass drill is the best way to plant seed into existing sod or firmly packed bare dirt. Grass drills with no-till attachments can plant seed into grass sod without any pretillage (fig. 5-3). Reduced soil erosion and fewer weeds are advantages of no-till drilling into sod. Grass drills work best if the soil and the vegetation are dry and most of the thatch and standing dead material are removed by burn-

Fig. 5-1. Broadcast-seeding with a Vicon fertilizer spreader and dragging a piece of fence to incorporate the seed into the soil. Photo by David O'Shields.

Fig. 5-2. Seeding natives with a Brillion grass seeder. Seed is dropped in between two steel wheel gangs; one conditions the soil, and the other cultipacks the seed. Photo by David O'Shields.

ing or haying. When operating properly, a grass drill with a no-till attachment moves the thatch with trash plows, cuts a shallow furrow, meters the seed at the selected rate, plants the seed ⅛ to ¼ inch deep, and presses the seed into the soil. In some areas, grass drills can be rented from government agencies. Check with your local Natural Resources Conservation Service for information on renting a grass drill. Note: A grass drill is a very specialized piece of equipment and should be operated by an experienced person.

To achieve the best performance and outcome with a grass drill, the seed must be properly mixed and calibrated, and the drill must be operated correctly. The following are some best practices to optimize the use of a grass drill in planting prairie seed.

Assign each species to the appropriate box (fig. 5-4). Seed size and the extent to which the seed has been cleaned will determine which seed box to use. Seed with beards and awns removed (debearded and deawned) will flow through the drill much faster than seed with those parts intact.

Only small seed and deawned and debearded seed should go into the front small seed box. This will include the seed of most forbs (except *Silphium* species) and some grasses. Seed flow can be impeded and uneven seeding will result if larger seeds are put into the front small seed box (table 5-1).

Fig. 5-3. Seeding natives with a Truax FLEXII grass and grain drill. No tillage is necessary prior to seeding, because this drill is designed to plant the seed into existing sod. Photo by David O'Shields.

Fig. 5-4. Front small seed box on a Truax FLEXII grass and grain drill. The small fluted wheel meters small seed through the drill. Photo by David O'Shields.

The middle seed box is reserved for fluffy seed with beards and awns intact (fig. 5-5, table 5-2). Aggressive feed wheels in the middle box are designed to pull apart fluffy seed. If debearded seed, deawned seed, and any other seed that can be poured are put into the middle fluffy seed box, the seed will rapidly flow through the drill and result in uneven seeding.

The rear cool-season grass and grain box is reserved for large, flowable seed. Debearded and deawned grass seed like Canada wild rye, Indian grass, big bluestem, and little bluestem and large forb seed like compass plant and rosinweed should be put into the rear cool-season grass and grain box (fig. 5-6, table 5-2).

Consider broadcast-seeding (by hand or seeder) the very small seed. Some practitioners will hand-broadcast very small seed (100,000 or more seeds per ounce) instead of using the grass drill (table 5-2). It is thought that a grass drill plants very small seed too deep. This may work well for smaller sites. However, hand-seeding and getting an even coverage of seed in a large planting may not be possible or practical. In this case, we recommend mixing all the very small-seeded species (table 5-2) together with an equal amount of scoopable kitty litter. Remove one or two discharge tubes from the front small seed box on the grass drill, and add the very small seed mix in the well(s) where the tubes were removed. Seed will randomly fall to the soil surface and will likely get pressed into the soil by the drill and tractor tires as the units pass over.

Mix the seed. The volume of seed in the drill boxes can affect the seeding

Table 5-1. Small-seeded Prairie Species That Can Be Broadcast onto the Surface without Being Incorporated into the Soil

Species	Moisture Class	Seeds/Ounce
Grasses/sedges		
Bluejoint grass	W-M	248,880
Brown fox sedge	W-M	100,000
Fowl manna grass	W-M	160,000
June grass	D	400,000
Forbs		
White sage	M-D	250,000
Heath aster	M-D	200,000
Silky aster	D	476,000
Harebell	D	900,000
Bottle gentian	W-M	280,000
Sneezeweed	W-M	130,000
Great St. John's wort	W-M	190,000
Great blue lobelia	W-M	500,000
Foxglove beardtongue	M	130,000
Common mountain mint	W-M	220,000
Grass-leaved goldenrod	W-M	200,000
Old field goldenrod	D	300,000
Showy goldenrod	M-D	103,600
Culver's root	W-M	800,000

rate. If the seed level drops below the agitators in the seed boxes, seed doesn't feed as efficiently, resulting in uneven seeding. For smaller sites (less than 1 acre), the total quantity of seed may not adequately fill the seed boxes, and thus the agitators may not work properly. This can be corrected by adding inert material to the seed to increase the volume. Filler should be similar in size to the seed in the mixture. After mixing each species and assigning it to the appropriate seed box, add an equal part (by volume) of inert material. Add scoopable cat litter to the seed that is to go in the front small seed box. For seed in the fluffy seed box, add an equal part of vermiculite. For seed in the cool-season grass and grain box, add an equal part of cracked corn.

Drill the entire planting site twice. To insure that seed is evenly planted, drill the entire site twice, each time using half the seed. Note: To seed the entire site twice, the drill will have to be calibrated using only half the total seed mix. If possible, we recommend that the site be drilled at a right angle to the first seed-

Fig. 5-5. Middle fluffy seed box on a no-till drill. The augers and picker wheels are designed to pull apart prairie seed that have their beards and awns intact. Photo by David O'Shields.

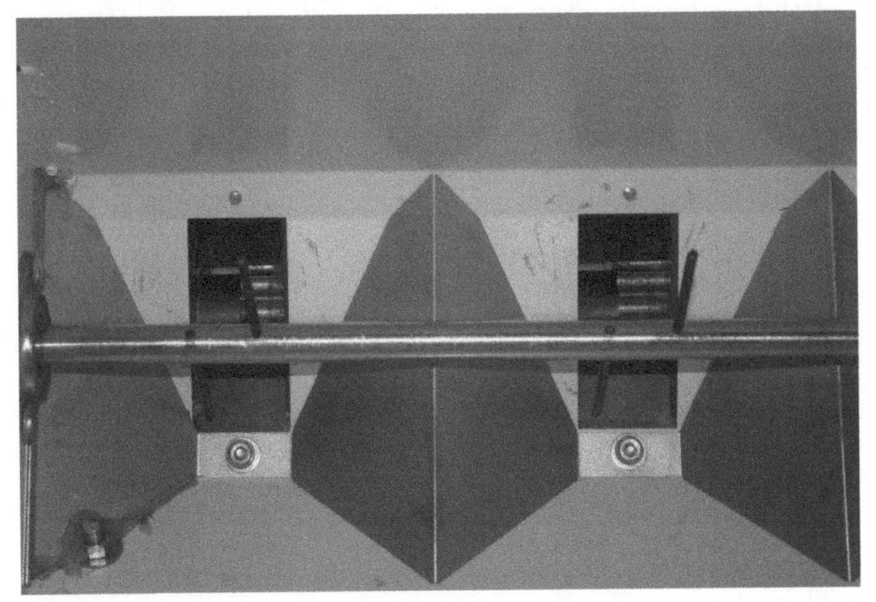

Fig. 5-6. Rear cool-season grass and grain seed box. The fluted picker wheels are designed to handle seed that have their beards and awns removed. Photo by David O'Shields.

Table 5-2. Seed Drill Box Designations of Selected Prairie Species

Species	Moisture Class	Seeds/Ounce	Awns, Beards, Hulls	
			Removed	Intact
Grasses				
Big bluestem	W-M-D	10,000	rear box	middle box
Side-oats grama	M-D	8,650		middle box
Kalm's bromegrass	M	8,000		rear box
Bluejoint grass	W-M	248,880	broadcast	middle box
Bicknell's sedge	M	17,000	front box	
Shortbeak sedge	M-D	29,000	front box	
Heavy sedge	M-D	12,000	front box	
Troublesome sedge	M-D	25,000	front box	
Broom sedge	W-M	84,000	front box	
Awlfruit sedge	W-M	34,000	front box	
Brown fox sedge	W-M	100,000	broadcast	
Canada wild rye	M-D	6,200	rear box	middle box
Virginia wild rye	W-M	4,200	rear box	middle box
Fowl manna grass	W-M	160,000	broadcast	
June grass	D	400,000	broadcast	
Switchgrass	W-M-D	16,000	front box	
Little bluestem	M-D	15,300	rear box	middle box
Indian grass	M-D	11,500	rear box	middle box
Prairie cord grass	W-M	6,040	rear box	middle box
Tall dropseed	M-D	30,000	front box	
Prairie dropseed	M-D	15,000	front box	
Porcupine grass	D	2,132	rear box	middle box
Forbs				
Wild garlic	W-M-D	8,398		rear box
Leadplant	M-D	17,884	front box	rear box
Thimbleweed	M-D	16,485	front box	middle box
Tall thimbleweed	M-D	28,000	front box	middle box
White sage	M-D	250,000	broadcast	
Swamp milkweed	W-M	4,800	middle box	rear box
Butterfly milkweed	M-D	3,350	middle box	rear box
Whorled milkweed	M-D	11,000	middle box	rear box
Upland white aster	D	64,000	front box	middle box
Milk vetch	W-M-D	18,662	front box	
White wild indigo	M-D	1,700	rear box	
Harebell	D	900,000	broadcast	
Partridge pea	M-D	2,700	front box	
Prairie coreopsis	M-D	11,000	rear box	
White prairie clover	M-D	18,230	front box	rear box

Table 5-2. (*Continued*)

Species	Moisture Class	Seeds/Ounce	Awns, Beards, Hulls	
			Removed	Intact
Purple prairie clover	M-D	18,950	front box	rear box
Showy tick trefoil	M	5,500	front box	middle box
Pale purple coneflower	M-D	5,300		rear box
Rattlesnake master	W-M-D	7,500		rear box
Joe Pye weed	W-M	95,000	front box	middle box
Tall boneset	M-D	50,000	front box	middle box
Grass-leaved goldenrod	W-M	200,000	broadcast	middle box
Bottle gentian	W-M	280,000	broadcast	
Sneezeweed	W-M	130,000	broadcast	middle box
Saw-tooth sunflower	W-M-D	13,440	front box	
Western sunflower	M-D	14,000	front box	
Prairie sunflower	M-D	4,480	front box	
Ox-eye sunflower	M-D	6,300	front box	
Great St. John's wort	W-M	190,000	broadcast	
Round-headed bush clover	M-D	9,000	front box	rear box
Rough blazing star	M-D	15,500	front box	middle box
Meadow blazing star	W-M	11,760	front box	middle box
Prairie blazing star	M	11,000	front box	middle box
Michigan lily	W-M	10,000		rear box
Great blue lobelia	W-M	500,000	broadcast	
Wild lupine	D	1,100	front box	
Wild bergamot	W-M-D	75,000	front box	
Dotted mint	D	90,000	front box	
Stiff goldenrod	M-D	41,000	front box	middle box
Wild quinine	W-M-D	7,000	front box	rear box
Foxglove beardtongue	M	130,000	broadcast	
Large-flowered beardtongue	D	14,000	front box	
Prairie phlox	M-D	19,000	front box	
Common mountain mint	W-M	220,000	broadcast	
Prairie buttercup	D	20,000	front box	
Gray-headed coneflower	M-D	30,000	front box	
Wild rose	M-D	500		rear box
Black-eyed Susan	M-D	92,000	front box	

Table 5-2. (*Continued*)

Species	Moisture Class	Seeds/Ounce	Awns, Beards, Hulls Removed	Awns, Beards, Hulls Intact
Fragrant coneflower	W-M	43,000	front box	
Wild petunia	M-D	5,200	front box	
Rosinweed	W-M-D	1,200		rear box
Compass plant	M-D	660		rear box
Old field goldenrod	D	300,000	broadcast	middle box
Showy goldenrod	M-D	103,600	broadcast	middle box
Heath aster	M-D	200,000	broadcast	middle box
Smooth blue aster	M-D	52,670	front box	middle box
New England aster	W-M-D	67,500	front box	middle box
Silky aster	D	476,000	broadcast	middle box
Purple meadow-rue	W-M	11,000	front box	
Prairie spiderwort	M-D	8,000	front box	
Ohio spiderwort	W-M-D	8,000	front box	
Blue vervain	W-M	93,000	front box	
Hoary vervain	M-D	28,000	front box	
Ironweed	W-M	24,000	front box	middle box
Culver's root	W-M	800,000	broadcast	
Heartleaf alexanders	M-D	12,000	front box	
Golden alexanders	W-M	11,000	front box	

ing pass. This will minimize potential bare spots (the area not drilled) in the planting site and will result in a more randomly planted prairie and thus less of a row look.

Calibrate each box separately. Each box has an independent adjustment to set the seeding rate. The front and rear boxes on the grass drill have gate levers that control the seeding rate. The gates should be opened just enough to allow the largest seed to pass through the fluted picker wheels; otherwise, large seed could be damaged. Adjusting the seeding rate on the middle fluffy seed box involves moving a roller chain from one set of sprockets to another. Each set of sprockets corresponds to a relative output of seed. For step-by-step instructions on how to calibrate a grass drill, see the appendix at the end of this chapter.

Always operate a grass drill at the recommended ground speed. The ground speed recommended for grass drills is 4 to 5 miles per hour (Truax 2004). If the grass drill is equipped with no-till trash plows, ground speed should be reduced by one-third (Truax 2004). Excessive ground speed will cause the drill to plant the seed improperly.

Adjust the drill when operating it. Trash plows should only scratch the soil surface as they move thatch away from the seed furrow. Seed will be planted too deep if the trash plows dig into the ground. Even when adjusted properly, the trash plows can dig into the ground when drilling on uneven ground such as in roadside ditches. For this reason, some practitioners will remove the no-till trash plows from the drill when seeding.

Inspect the drill while operating it. Inspect the depth bands frequently while drilling. Avoid drilling in wet conditions. Mud build-up on the depth bands can change the seeding depth. A stiff putty knife works well to remove mud from the depth bands. Feeder tubes from the seed boxes can plug. Periodically, squeeze and shake the feeder tubes connected to the fluffy and cool-season boxes. Individual compartments within the small seed box should have similar quantities of seed remaining while drilling. A compartment with more seed than another compartment may indicate a plugged feeder tube. Replacing the black feeder tubes with clear tubing on the small seed box allows the operator to visually inspect for plugged tubes while seeding.

HYDROSEEDING

In this unique seeding method, seed is mixed with water, mulch, and tackifier — a binding or adhesive agent that holds the fiber mulch in place when applied over the seedbed — to form a slurry that is sprayed directly on the ground. Many county road departments and some landscaping companies use hydroseeding to seed prairie (see chapter 12). While this method of seeding is restricted to professionals, you may decide to hire a local company to hydroseed your prairie. We recommend that your seeding contractor hydroseed with a two-step process. The first step is to broadcast the seed (see the Broadcast Seeding subsection in this chapter). The second step is to spray the hydromulch slurry (without seed) over the seeded area. This two-step process will help insure that the seed is not suspended in the mulch, where it can desiccate. Additives can be included in the slurry to reduce soil erosion (Meyermann 2008).

SEEDING DESIGN

Shotgun seeding employs one seed mix for planting on sites with varying soil moisture conditions. The seed mix should contain species that match each soil moisture condition of the site. This seeding design may not maximize establishment of all species in the mix but requires less time to plant than the mosaic seeding design described below.

Mosaic seeding involves using multiple seed mixes that are sowed separately. Mosaic seeding can be applied to sites that have varying soil types, hydrology, slope, and aspect. Species with specific environmental preferences can be seeded into those areas where they will most likely establish and persist. Dry-adapted species are seeded on dry sites, and wet-adapted species are seeded on wet sites. The benefits of mosaic seeding are economic efficiency, maximum establishment of all species in the seed mix, and a more natural-looking prairie reconstruction.

Summary

- Planting natives at the appropriate time will increase the chance of a successful planting.
- Dormant seeding will permit stratification of the forb seed and improve germination.
- Broadcast seeding is not recommended for dormant seeding.
- We recommend dormant seeding if the seed mix contains 50 percent or more forb seed.
- Grass seed in the mix should be increased by 25 percent if dormant-seeded to compensate for seed mortality.
- Early spring seeding favors cool-season grasses and sedges and some forbs.
- Late spring planting favors germination of warm-season grasses and some forbs.
- Artificially stratified forb seed can be used for a spring planting if there is adequate soil moisture at the time of seeding and the seed is planted ⅛ inch deep and cultipacked.
- Seeding natives in mid to late summer is not recommended.
- If the seed is frost-seeded with the broadcast method, the seeding rate should be increased by 25 percent to compensate for seed loss due to wind erosion and predation.
- Drag, harrow, or rake broadcast seed until it disappears in the soil, followed by cultipacking, to maximize seed-to-soil contact.
- Seed planted too deep will not emerge.
- Adjustment and calibration of the grass drill are essential for proper planting depth and metering of the seed.
- The extent to which the seed of a species is cleaned and its size will determine the appropriate seed box in the grass drill.
- Adding inert material (vermiculite, kitty litter, and cracked corn) to increase the volume of seed may be needed to improve metering of seed when drilling sites are less than 1 acre.

- To insure that seed is evenly planted, seed the site twice, using half the seed for each pass.
- Periodically check the grass drill feeder tubes for plugging during seeding.
- Avoid drilling in wet conditions.
- Have your hydroseeding contractor use the two-step process.
- Mosaic seeding can maximize establishment of all species in the seed mix and create a more natural-looking prairie.

Appendix. Calibration Instructions for a Grass Drill When Seeding Mixed Species

1. Attach the drill to a tractor, and park on a level surface. Lift the drill with the hydraulics. Set the parking brake, and shut off the engine.
2. Place a jack stand under the drive wheel to keep it elevated off the ground.
3. Slowly lower the hydraulics so the drill frame rests on the jack stand.
4. Block the nondrive wheel (both front and rear).
5. Engage the drill clutch. Turn the feed wheel clockwise, and check to insure that all feeding mechanisms are turning.
6. Measure (in feet) the outside circumference of the drive wheel.
7. Measure the effective drilling width. This can be done by measuring the distance from the planting wheel on one end to the planting wheel on the other end.
8. Count how many discharge tubes are mounted to the box being calibrated.
9. Weigh the total amount of seed that is to go into the box being calibrated.
10. Mark the drive tire sidewall with a tire crayon.
11. Disconnect one feed hose from the well of the box to be calibrated.
12. Place a bucket under the disconnected well to catch the seed.
13. Add seed mix only to the well of the box with the disconnected hose. To insure that there is enough seed for adequate calibration, fill seed to the top of the auger wheel in the middle box, or fill the front or rear box until the picker wheel disappears.
14. Spin the drive wheel for 5 revolutions to bleed the feeder chutes.
15. Empty the seed caught in the bucket into the well, and place the bucket back under the well.
16. Turn the drive wheel for 10 revolutions.
17. Weigh (in ounces) the discharge seed caught in the bucket.
18. Empty the seed caught in the bucket into the well, and place the bucket back under the well.

19. To calibrate the seeding rate for the box (bulk pounds for your planting site), follow steps 20 through 23.
20. Multiply the weight of the discharged seed by the total number of discharge tubes on the box. This will give you the weight (in ounces) of seed discharged from all tubes in 10 revolutions.
21. Multiply 10 by the circumference of the drive wheel. This will give you the total area drilled with 10 revolutions.
22. Divide the total area of the planting site (measured in square feet) by the value calculated in step 21. This will give you the number of seeding passes required to seed the entire planting site.
23. Multiply the number of seeding passes by the value calculated in step 20 and divide by 16. This value is the total bulk weight (in pounds) of seed required for the planting site.
24. Adjust the feed gate as needed and recalibrate, starting at step 15.
25. Proceed to the next box for calibration, starting at step 8.

Source: Parts of this appendix were excerpted from the Truax FLEXII operator's manual (2004).

First-Season Management

DAVE WILLIAMS

Overview

The establishment of a prairie plant community takes 3 to 5 years. Without early management of the vegetation during this critical time, weeds and woody plants will displace the emerging and newly established native plants, resulting in a weedy plant community that will persist for many years. The goal after seeding is to reduce unwanted plants — most commonly weeds — and stimulate establishment and growth of the native plants until the prairie plant community is established.

This chapter will discuss a variety of early management techniques, including frequent mowing, herbicide use, manual maintenance (hand-pulling, hoeing, and girdling), prescribed burning, and irrigation to control unwanted plants and enhance the natives.

Prairie Plant Establishment and Weeds

Weed control in a newly seeded native planting should be a high priority in early reconstruction management. Fast-growing annual weeds can form a closed canopy over native perennial seedlings in less than 30 days, reducing light to a fraction of full sunlight. Low light intensity stunts native seedlings' development, making them susceptible to winter mortality (Williams et al. 2007).

Perennial weeds can also negatively affect native plant establishment. Of most concern to resource managers are the perennial weeds that displace native plants and invade established prairie plantings. Perennial weeds and perennial prairie plants share many similar traits. Both produce seed, spread vegetatively, and occupy the same root zone and above-ground space. Some perennial weeds such as Canada thistle and leafy spurge form dense colonies that eliminate native plants (Butler et al. 2004). Methods to control perennial weeds must be used carefully because they will also have the same effect on the native perennials.

Mowing

Weeds that are allowed to grow high enough to create a closed canopy during the first few years of a prairie planting will reduce germination, growth, and survival of the perennial prairie plants (Williams et al. 2007). This can create long-term maintenance problems. Frequent mowing is an effective technique to prevent a weed canopy from forming in a new prairie planting. Mowing can be done with any type of mower as long as the mower deck can be raised at least 4 inches. Some practitioners prefer using a flail-type mower because the biomass is cut into smaller pieces and does not leave a windrow (a thick layer of thatch) on the surface.

As a general rule of thumb, do not let weeds and other vegetation get taller than knee high in the first growing season. Mow to a height of 4 to 6 inches whenever the vegetation grows 12 to 18 inches high in the first growing season (fig. 6-1). Don't be concerned about damaging the natives by mowing. Most prairie seedlings will grow below the 4- to 6-inch mow height in the first growing season. The frequency and duration of mowing depend upon the weed density and climate conditions during the growing season. Typically in Iowa, with average precipitation, mowing may be needed every 3 weeks from early May to early September in the first growing season. This frequent mowing regime will curtail the growth and seed set of weeds while preventing thatch build-up that can smother native seedlings.

Mowing in the second growing season depends upon the density of persistent perennial and biennial weeds. To avoid damaging the native plants, mowing height should never go below 12 inches in the second growing season. Time between mowing can be monthly or longer depending upon the weed pressure. For scattered weed patches, consider spot mowing or hand-pulling to minimize the impact upon developing prairie plants. If there is a flush of tall, rank biennial weeds like Queen Anne's lace, sweet clovers, or wild parsnip, it is important to mow or pull just prior to flowering; this will severely curtail or eliminate the plants' ability to flower and go to seed (Sheley 2001).

By the third growing season, most of the vegetative growth throughout the site should be prairie plants, and mowing should not be needed. If there is a threat of a weed canopy in year 3, a stand evaluation should be conducted to determine if adequate numbers of prairie plants remain (see chapter 7). If native plant establishment is less than 1 plant per square foot, we recommend using

Fig. 6-1. Frequent mowing in the first year of a prairie planting removes the canopy of fast-growing annual weeds, increasing sunlight and promoting growth of the slower-growing prairie perennials. Photo by David O'Shields.

a stand enhancement technique to add more prairie grasses and forbs to the stand (see chapter 5).

Mowing is only partially effective at controlling persistent perennial weeds and woody plants. It will eliminate seed production and reduce the weed canopy if implemented at the right time during the growing season, but it will have little effect on rhizomal spread — in some cases it may increase it (Lalonde et al. 1994). In those circumstances, herbicides may be needed to control persistent perennial weeds.

Herbicides

Herbicides, when used carefully at rates listed on the label, can be very effective at controlling persistent perennial weeds and woody plants (see table 4-1). Careless application will result in killing native species. Just how many weeds are considered weedy in a native planting is a matter of personal preference, but addressing weed issues early can save heartache later on. There will be less damage to native plants if chemical control is used within the first few years of a seeding, when weeds are less abundant and can be spot-sprayed. Waiting until weeds are abundant in the planting can turn spot spraying into blanket spraying, which is extremely damaging to natives. States require certification testing and licensing to purchase and apply certain pesticides. Contact your state's Department of Agriculture to obtain more information on pesticide certification. Always read and follow label directions. The following are some strategies to minimize damage to native plants when using herbicides to control unwanted plants.

Spray only the persistent perennial weeds and woody plants (see table 4-1). Over time, prairie plants will exclude most other weeds from the planting.

Spray when the natives are dormant. Crown vetch and leafy spurge remain green into fall and can be sprayed after most native plants are dormant.

Use herbicides that are species-specific. Some herbicides work better than others on individual weed and woody species. Clopyralid or chlorsulfuron is more effective at controlling Canada thistle than glyphosate. Fosamine ammonium can be foliar-sprayed on woody plants without affecting native forbs (nonwoody) and grasses.

Spray the weeds at the proper stage of plant development. The label will indicate at what stage of development the weed species is most susceptible to the effects of the chemical. Rosette to bud, up to 5 leaf stage, 1 to 3 leaf stage before vining, and boot to early seed-head stage are some examples of specific label recommendations for optimum spraying times.

Apply herbicides at the rate specified by the label. The herbicide application rate will vary according to the weed species and the severity of infestation.

Use spot spraying. To minimize overspraying onto nontarget plants, use a hand wand instead of a boom sprayer. A backpack sprayer with a spray wand extension allows the operator to place the nozzle tip very close to the weed to minimize overspraying (fig. 6-2).

Use a boom sprayer only on large, dense weed patches.

Avoid creating drift when spraying. Lower the spray pressure and increase the size of the nozzle orifice to reduce spray drift. Don't spray on windy days. Consider spraying in the early morning or early evening when the wind tends to be calm.

Cut rather than foliar-spray woody plants. Because many brush herbicides require complete coverage when foliar-sprayed, there is the potential for excessive overspraying onto nontarget plants. A cut-stump herbicide to prevent the stump from resprouting can be applied precisely where it is needed without damaging the surrounding vegetation (fig. 6-3, table 4-2).

Do not apply an herbicide to a cut stump that is actively flowing with sap. Sap flow will cause the herbicide to run off the cut stump into the soil and kill nearby vegetation. This is often referred to as the ring of death.

Fig. 6-2. Spraying Canada thistle in a prairie planting. Using a wand extension on a backpacker sprayer can greatly reduce overspray on nontarget plants. Photo by David O'Shields.

Fig. 6-3. To reduce the risk of soil contamination and damage to nontarget plants, apply a cut-stump herbicide to the cambium layer of the woody plant. Photo by David O'Shields.

Manual Maintenance

There are nonherbicide methods to control weeds and woody plants in a native planting. These methods require extra physical exertion and time but can be the least damaging to the surrounding native plants. The severity of the infestation and the stamina of the land manager will dictate whether these methods are practical.

The best time to hand-weed is immediately after a rainstorm, when the ground is soft and a large portion of the root can be extracted. Perennial rhizomatous weeds like Canada thistle and leafy spurge will require several pullings in the same year and may require weeding for 2 or more years. Wear thick gloves—Canada thistle plants are prickly, and leafy spurge sap can cause dermal reactions.

Chopping with a hoe or spade or cutting with shears works very well on biennial plants. Cutting the plant under the soil surface or near its base as it begins to flower will greatly reduce its ability to regrow and produce seed.

A heavy-duty string trimmer fitted with a steel brush blade or plastic knives can selectively cut small weed patches and smaller-diameter woody plants scattered throughout a planting (fig. 6-4). This piece of equipment can be dangerous to operate. Always wear proper safety equipment: gloves, a long-sleeved shirt,

Fig. 6-4. Invasion of gray dogwood into a planted prairie. A brush blade fastened to a string trimmer can be an effective way to remove brush. Photo by David O'Shields.

safety glasses, hearing protection, a hard hat, chaps, and steel-toed shoes. Read and follow the recommendations in the owner's manual for safe operation.

Girdling (like rabbits do to young trees and shrubs in the winter) can kill woody plants. Girdling is accomplished by scraping a thin layer of bark off the stem all the way around the plant near its base. Immediately inside the outer layer of bark is a very thin green layer (phloem) that must be removed. On young woody plants, the tissue can be scraped off fairly easily with a sharp pocket knife. Be careful not to girdle too deeply. Cutting too deeply will stimulate the plant to produce new sucker shoots, the same response when the stem is completely severed. Girdling stops nutrients generated in the leaves from moving into the roots, and the roots starve and die. Be sure to girdle any new sucker shoots. Any shoots not girdled will allow the plant to survive.

Always wear protective clothing and gloves to guard against thorns and plant compounds that can cause severe skin reactions.

Prescribed Burning

A prairie planting should be burned as soon as the site can carry a continuous fire. Proper use of prescribed fire will accelerate the growth of most prairie plants and deter cool-season weeds and small woody plants. Typically, there is not enough fine fuel (grass leaves) to carry a fire in a 1- or 2-year-old planted prairie due to frequent mowing. By the end of the third growing season, however, there should be enough grass growth to carry a fire. The first prescribed fire on a newly reconstructed prairie is often done in the spring to stimulate the warm-season prairie grasses. The management objectives should determine the frequency and timing of prescribed fires for subsequent prescribed burning (see chapter 10). Prescribed burning should be done only by trained and experienced personnel. To learn more about prescribed burning, visit the Iowa Natural Resources Conservation Service website at http://www.nrcs.usda.gov/.

Irrigation

Irrigation can be an important management tool. Once a seed germinates, there is a critical phase of development between emergence and the time that the seedling develops its first true leaves. While the plant is still in this cotyledon stage, it cannot survive an extended period of drought. If rainfall is not adequate, seedlings will benefit from being watered 1 to 2 inches every 3 days during the first growing season (Morgan 1995). Irrigation increases the probability that the plants will survive into the second growing season and beyond.

Fertilizers

Fertilizers are not recommended for native plantings. Most plants including natives benefit from fertilizers, but weeds benefit *more*, making fertilizers a poor management strategy. Fertilizing a newly planted prairie will disproportionately favor opportunistic weedy species. Most native species are well adapted for nutrient-poor soil. In some cases, fertilization may also damage or kill native seedlings (see chapter 16).

Summary

- Low light levels in a closed canopy reduce emergence and growth of native seedlings in year 1 (Williams et al. 2007).
- Cool, moist conditions created by a closed canopy encourage pathogen growth that can injure or kill prairie seed and seedlings.
- A closed canopy creates cover for herbivores that will injure prairie seedlings in year 1.
- Mow to a height of 4 to 6 inches whenever the vegetation grows 12 to 18 inches high in the first growing season.
- Frequent mowing will prevent the build-up of thatch that can smother native seedlings and curtail the growth and seed set of weeds.
- Mowing is not required in year 2 unless there is a resurgence of persistent perennial or biennial weeds.
- To avoid damaging native plants, mowing height should never go below 12 inches in the second year.
- For scattered weed patches, consider spot mowing or hand-weeding to minimize the impact on developing prairie plants.
- Frequent mowing may not control perennial weeds and woody plants.
- Manual maintenance can reduce problem perennial weeds if done multiple times during the growing season. Follow-up treatments in subsequent years may be needed to control perennial weeds.
- Spraying persistent perennial weeds when the natives are dormant can minimize herbicide damage to nontarget species.
- Native plantings near woods or brushy areas are at the highest risk for woody plant invasion by seed or clonal spread.
- Herbicides applied to the cut stumps of woody plants can be very effective at killing the plant.
- Do not apply herbicides if sap is flowing from the cut stump.

- Girdling (without applying herbicides) can control some woody plants. Any subsequent resprouting needs to be girdled to kill the plant.
- Girdling too deeply stimulates suckering.
- Prescribed fire will accelerate the growth and reproduction of most prairie plants.
- If rainfall is not adequate, smaller plantings should be watered 1 to 2 inches every 3 days during the first growing season (Morgan 1995).
- Fertilizers are not recommended for prairie reconstructions.

7 Evaluating Stand Establishment and Seedling Identification

DAVE WILLIAMS

Overview

This chapter is intended to assist the practitioner-landowner in sampling and evaluating prairie plant establishment in a new seeding. Deciding where to sample, how many samples to take, what to measure, and how to analyze the data for an assessment of prairie plant establishment are discussed. This chapter outlines techniques for developing seedling identification skills to insure an accurate assessment of the initial establishment of prairie vegetation. Information on assessing prairie remnants is discussed in chapter 8.

Sampling to Assess Prairie Plant Establishment

Assessing the establishment of prairie plants in the first or second growing season can eliminate reseeding a successful planting unnecessarily or prolonging the maintenance of a failed planting. Establishment often varies throughout a planting. Variations in topography and soil type expose the seed to different growing conditions, sometimes resulting in poor emergence. One purpose of vegetative sampling is to find areas within the planting that have too few seedlings, so they can be reseeded. Sampling can also detect areas where there are persistent perennial weeds, which will reduce prairie plant establishment (see chapter 6). Controlling these weeds early in the reconstruction can save the landowner time and money.

Sampling is a systematic process used to gather a small part (or sample) of something and analyze it to answer a basic question. A basic question asked by managers and landowners about a new prairie planting might be, Are there enough prairie plants in the planting? To answer this question, you can proceed in one of two ways. First, you could identify and count every prairie plant in the planting. Then take that number and divide it by the total square feet in the planting, which will result in the number of seedlings per square foot. This number can be compared to the recommended number of seedlings per square foot that are needed for adequate native plant establishment. The first method would be extremely time-consuming, but you could accurately calculate prairie plant establishment for the planting. Second, you could choose many different locations throughout the planting, and identify and count only the prairie plants that occur in a very small area, 1 foot square, at each location. Prairie plant establishment could then be calculated by adding up all prairie plants found, then dividing the total by the total square feet sampled. This number, as with the number in the first method, can be compared to the recommended number of seedlings that are needed for adequate native plant establishment. Clearly, the second method is easier and saves time. If the second method is correctly done, the number of plants per square foot should be very similar to number obtained from the first method. Sampling is an excellent assessment tool in prairie management.

How Much and Where to Sample

Determining how much vegetative sampling is needed depends upon the complexity of the landscape. For planting sites, regardless of size, that don't have much variation in topography and soil type, a minimum of 20 to 30 samples will be needed to assess prairie plant establishment (Witmer 1999). In plantings that have varying habitats (such as varying slopes and aspects, rock outcrops, swales, or waterways), additional vegetative sampling is required. To accurately assess seedling establishment in plantings with a variety of habitats, areas of the site with similar environments should be sampled and analyzed separately. This is called stratified sampling. Stratified sampling requires dividing the site into habitat types based on environment and calculating prairie plant establishment for each habitat. We recommend that you take a minimum of 20 to 30 samples for each habitat type. An advantage of stratified sampling is that areas in the planting with poor seedling establishment that may otherwise go undetected can be identified.

It is human nature to choose sampling locations that contain only a few plants, because this makes identification and counts much easier, but the results will not provide the accurate information needed to successfully manage the planting. Instead, vegetation should be sampled at random locations within habitats to obtain the most representative information about species composition. This process is called randomized sampling. The following steps will help insure that sampling is randomized *before* you go into the field.

Review the site map, and mark each distinct habitat type that should be sampled and analyzed separately.

On the map, select a starting point anywhere along the boundary of each habitat type to be sampled. Choose an end point on the opposite side of the habitat type that is farthest away from the starting point. With a pencil, connect the points. This line is called a transect. Measure the transect length using the map scale. Divide the number of samples (20 to 30) by the transect length. This will give you the distance between each sample to be taken along the transect in the field.

Measure your pace distance, using a normal walking speed.

Divide the distance between samples by your pace distance to determine how many paces are needed between samples.

Collecting Vegetation Samples

Vegetation sampling in grasslands is often done using a quadrat, a small frame with a known area measurement inside the frame. Seedling density and frequency can be accurately measured using a 1-foot-square quadrat frame (Dayton 1988). Quadrats can be built from flexible PVC tubing, wood, or wire (figs. 7-1 and 7-2). Most grassland quadrats are three-sided, with one side left open for easy insertion into the sample area. The frame is inserted near the base of the plants at preselected locations along the transect, and the vegetation inside the frame is identified and counted. One sample equals one quadrat of vegetation. The following steps should be used when entering the field to sample.

Using the site map as a guide, locate the transect line's starting and end points, and mark them with field flags.

Starting at one of the field flags, walk a straight line toward the other flag, using the number of paces needed between samples.

Place the quadrat at your feet, and sample the vegetation.

Continue along the transect, walking the calculated number of paces and taking a quadrat sample, until all the samples are taken.

For ease of seedling identification, the best time to sample a new planting is late August to early September. By the end of summer, most prairie seedlings will have grown enough to be accurately identified.

Assessing Stand Establishment Using Plant Density

There are several ways to sample vegetation within the quadrat frame. A good way to assess prairie plant establishment in a newly reconstructed prairie is by

Fig. 7-1. Circular quadrats made from PVC tubing. Sampling area is 2.69 square feet or .25 square meter in the large quadrat and .93 square foot or .10 square meter in the small quadrat. Photo by David O'Shields.

Fig. 7-2. Open-ended quadrats can be inserted easily into dense vegetation. Sampling area inside each quadrat frame is .93 square foot or .10 square meter. Photo by David O'Shields.

measuring plant density. This involves identifying and counting each prairie plant within the quadrat frame. Plant density is an excellent sampling method in early reconstructions (year 1), because prairie seedlings have not yet spread by rhizomes and/or produced multiple tillers and stems. Counting individual plants and stems is feasible. A native planting should have a minimum of 1 prairie plant per square foot (Morgan 1995). A planting that has less than 1 prairie plant per square foot by the end of year 2 is susceptible to weed invasion and may require additional management to control weeds (see chapter 6).

Developing a good data sheet is critical to any sampling method. Use a spreadsheet to create a data sheet for the field. The data sheet should be identical to the spreadsheet on your computer. This will reduce mistakes when entering data from the field into the computer. Organize the data sheet by rows and columns. Each column represents a quadrat sample, and each row represents a plant species. Arrange the species first by grasses, then by forbs and weeds. List all the native species seeded on the data sheet. They should be listed by the number of seeds per square foot planted (highest to lowest). When data are entered in the computer spreadsheet, enter the number of hash marks. Recording the data is easier when the highest-seeded species are clustered together on the data sheet. Record only the presence of persistent perennial weed species on the data sheet (table 7-1); it is not necessary to count their seedling numbers. Any persistent perennial weeds detected during quadrat sampling should trigger a scouting of the field to locate and map these plants for weed control (see the Scouting for Weeds section).

Place the quadrat at each random sample location, and record every prairie plant inside the frame. When identifying seedlings, follow the order listed on the data sheet; identify the native grasses first, followed by the native forbs, saving the weeds for last. Quadrat sampling with another person can be efficient. With a two-person sampling team, one person records the data, while the other person identifies the plants. Plants that are difficult to identify should be collected and bagged for later identification or flagged (label the flags A, B, C, etc.) and returned to in a couple of weeks. When collecting unknown species in bags, assign each plant a letter on its bag; record it on the data sheet so it can be changed on the sheet when identification is made.

Sampling is hard work. When the sun is beating down and the insects are biting, it is easy to get distracted and make mistakes. Periodically, check the data sheet to make sure that the quadrat column being recorded is the one that the seedlings are sampled from.

To calculate prairie plant establishment by density for the sample in table 7-1, follow the three steps listed below.

Table 7-1. Data Sheet from a 10-Acre Prairie Reconstruction

	Quadrat Samples (1 Square Foot Area)																			
Species	Q1	Q2	Q3	Q4	Q5	Q6	Q7	Q8	Q9	Q10	Q11	Q12	Q13	Q14	Q15	Q16	Q17	Q18	Q19	Q20
Big bluestem	✓		✓	✓✓			✓	✓										✓		
Indian grass	✓✓				✓	✓										✓				
Switchgrass	✓	✓	✓	✓✓	✓	✓	✓	✓												
Side-oats grama					✓	✓														✓
Black-eyed Susan	✓	✓✓		✓	✓	✓	✓					✓	✓							
Wild bergamot					✓			✓												
Stiff goldenrod			✓			✓	✓				✓									
Showy tick trefoil				✓					✓					✓✓✓						
Compass plant				✓																
Total seedlings	5	3	3	7	5	5	4	3	1	0	1	1	1	3	0	1	0	1	0	1
Canada thistle	P		P	P	P						P	P						P		
Smooth brome	P	P	P	P	P			P	P		P	P	P		P		P	P	P	

Note: P = present.

1. Sum the total prairie seedlings recorded in all quadrats:

   Total prairie seedlings = 5 + 3 + 3 + 7 + 5 + 5 + 4 + 3 + 1 + 0 + 1 + 1 + 1 + 3 + 0 + 1 + 0 + 1 + 0 + 1 = 45

2. Sum the total quadrat area sampled:

   Total sampling area = 20 (quadrat samples) × 1 ft^2 (quadrat area) = 20 ft^2

3. Divide the total prairie seedlings by the total sampling area:

   Prairie seedling establishment = 45 seedlings / 20 ft^2 = 2.3 seedlings/ft^2

Prairie seedling establishment can also be calculated for plant groups (grasses and forbs) by following the three steps, using the seedlings of the plant group of interest.

In this example, prairie plant establishment exceeds the minimum of 1 prairie plant per square foot, which is an adequate stand. In this example, the presence of Canada thistle should trigger the need to control this weed.

Determining plant frequency of a group of species can tell you how prominent they may be in the planting. In the case of forbs, a frequency of 40 percent or higher may indicate that the planting could develop into a forb-rich planting. In the case of prairie grasses, a frequency of 50 percent or more should result in an adequate stand (Rosburg 2006). To calculate grass and forb frequencies, follow the three steps listed below.

1. Sum the total number of quadrats where grasses and forbs were found:

   Grasses occurred in quadrats 1, 2, 3, 4, 5, 6, 7, 8, 16, 18, 20 = 11 quadrats
   Forbs occurred in quadrats 1, 2, 3, 4, 5, 6, 7, 8, 9, 11, 12, 13, 14 = 13 quadrats

2. Divide the number of quadrats with forbs and grasses by the total number of quadrats sampled:

   Grass frequency = 11 quadrats with grasses / 20 quadrats = 0.55
   Forb frequency = 13 quadrats with forbs / 20 quadrats = 0.65

3. Multiply the plant frequency by 100:

   Grass frequency (%) = 0.55 × 100 = 55%
   Forb frequency (%) = 0.65 × 100 = 65%

In this example, grasses have exceeded the 50 percent and are likely to become well established in the prairie planting. Due to the high forb frequency of 65 percent, this planting has the potential to develop into a forb-rich community.

Assessing Stand Establishment Using Species Frequency

Prairie plant establishment can also be assessed in a newly reconstructed prairie by calculating species frequency (Siefert and Rosburg 2004). This method uses the same randomized quadrat sampling technique described earlier. Prairie species are identified within the quadrat and recorded on the data sheet as present or absent (table 7-2, which uses the same data as table 7-1). Counting individual plants is not needed. An advantage of this method is that it doesn't require determining individual plants, which can be difficult in a more mature prairie reconstruction (2 or more years old) when many prairie species have produced multiple tillers and stems and have spread by rhizomes. A reconstructed prairie should have a prairie plant frequency of 50 percent or higher (Rosburg 2006). Any persistent perennial weeds detected during quadrat sampling should trigger a scouting of the field to locate and map these plants for weed control (see the Scouting for Weeds section).

To calculate prairie plant establishment by frequency, follow the three steps listed below.

1. Sum the number of quadrats where there was a prairie plant present:

   Prairie plants occurred in quadrats 1, 2, 3, 4, 5, 6, 7, 8, 9, 11, 12, 13, 14, 16, 18, 20 = 16 quadrats

2. Divide the total number of quadrats with prairie plants present by the total number of quadrats sampled:

   Plant frequency = 16 quadrats with prairie plants / 20 quadrats = 0.80

3. Multiply the prairie plant frequency by 100:

   Prairie plant frequency (%) = 0.80 × 100 = 80%

An estimate of the minimum total density can be made by letting P represent 1 (one prairie plant). In this example, 39 plants (total number of Ps) divided by 20 square feet (total sample area) = 1.95 plants per square foot. Here, prairie plant establishment exceeds the minimum of 1 prairie plant per square foot, or a minimum frequency of 50 percent, which is an adequate stand. Note that this is a minimum estimate; the real number of seedlings per square foot is likely to be higher. The presence of Canada thistle in this example should trigger the need to control this weed.

Table 7-2. Data Sheet from a 10-Acre Prairie Planting

	Quadrat Samples (1 Square Foot Area)																				
Species	Q1	Q2	Q3	Q4	Q5	Q6	Q7	Q8	Q9	Q10	Q11	Q12	Q13	Q14	Q15	Q16	Q17	Q18	Q19	Q20	FRQ
Big bluestem	P		P	P			P	P										P			0.30
Indian grass	P				P	P										P					0.20
Switchgrass	P	P	P	P	P	P	P	P													0.40
Side-oats grama					P	P														P	0.15
Black-eyed Susan	P	P		P	P	P	P					P	P								0.40
Wild bergamot					P			P													0.10
Stiff goldenrod			P			P	P				P										0.20
Showy tick trefoil				P					P					P							0.15
Compass plant				P																	0.05
Present/absent	P	P	P	P	P	P	P	P	P	A	P	P	P	P	A	P	A	P	A	P	0.80
Canada thistle	P		P	P	P						P	P									0.35
Smooth brome	P	P	P	P	P			P	P		P	P	P		P		P	P	P		0.70

Note: P = present, A = absent, FRQ = frequency.

Seedling Identification: The Challenge

The small size of seedlings makes their identification difficult. Perennial prairie plants typically grow slowly in the first year. Detecting identifying characteristics on small plants often requires a hand lens. Sometimes the only way to accurately access the characteristics of a plant is by destructive sampling—removing part of the seedling to crush a leaf, roll a stem, or examine a ligule under a magnifying glass. Destructive sampling should be done carefully, leaving as much of the seedling's root and leaf structure as possible.

Seedlings of most tallgrass prairie species do not look like their adult counterparts. Many species display a number of intermediate leaf shapes as they grow and mature. For example, the seedling leaf of compass plant is lance-shaped with a prominent midvein; the adult leaf is pinnately lobed. Seedling leaves of golden alexanders change from a kidney shape to a maple leaf shape to a compound trilobed leaf (Jackson and Dittmer 1997).

Seedlings and adult plants may also exhibit different colors. The hairs on an adult plant of white sage make it appear white, whereas the seedling—with very fine hairs—appears green. Likewise, at different times of the year, the changes in plant color can help with seedling identification. Foxtail turns yellow in September, making it easy to distinguish from native grass seedlings, which may remain a vivid green during this time.

Seedling identification requires using senses other than sight. Sometimes seedling characteristics cannot be seen with the naked eye. Leaf hairs on stiff goldenrod can be seen only using a hand lens, but when the leaf is rubbed between the fingers, the leaves feel very rough. This is due to the stiff hairs on the leaf surface. Leaves of some species produce an odor when crushed. Crushing purple prairie clover leaves produces a citrus odor. Likewise, crushing leaves of white sage produces a sage odor.

Look-alike weed seedlings add another level of complexity to identifying native seedling plants. Seedlings of many weed species resemble native seedlings. An example of this is seen in foxtail seedlings, which look strikingly like little bluestem. Little bluestem has a flattened tiller near the base of the plant that cannot be rolled between the fingers. The tiller base for foxtail is also flattened and cannot be rolled between the fingers. In this example, as with identifying many native seedlings, more than one characteristic needs to be used for identification. The presence of pubescence or hairs on the tiller can be a diagnostic characteristic. In this case, the tiller base of foxtail is hairless, whereas the tiller on little bluestem has coarse hairs.

Most prairie reconstructions in year 1 are very weedy. Finding small prairie seedlings among the weeds is difficult. Most prairie plants (except prairie annuals) seeded in late fall of the previous year to early summer of the current year will be 1 to 3 inches tall by early September of the first growing season. Sampling conducted in September should focus on plants that fit within this size range. This will help rule out a whole group of larger plants that are most likely annual or established perennial weeds. Commit to memory one or two key identification characteristics for each native species that you plant. This will help you narrow the search for prairie seedlings.

Seedling Identification: Key Characteristics

The following species are organized by similar key characteristics. When identifying a native seedling, look first for the key characteristics to determine its group; then find the species within the group that best fits the remaining identification characteristics. The following simple key is intended for prairie seedling identification in year 1. It utilizes the characteristics of 18 forbs and 7 grass species. By year 2, many prairie plants develop adult characteristics, limiting the value of some key characteristics listed in this section. For morphological characteristics for grasses and forbs, see figures 7-3 and 7-4.

GRASSES

Flattened tiller near base, auricles absent

- Big bluestem: small, ragged ligule; coarse hairs on tiller and leaf margins; rolled emergent leaf
- Little bluestem: tiller base so flat that it cannot be rolled between the fingers; small, ragged ligule; coarse hairs on tiller and leaf margins; flattened emergent leaf
- Indian grass: somewhat flattened tiller; tillers can be hairy or hairless; tall, stiff ligule that often can be seen without a hand lens; hairless leaf margins; rolled emergent leaf
- Side-oats grama: round, hairy tiller; small, ragged ligule; swollen glands on leaf margin hairs; rolled emergent leaf

Round tiller near base, auricles present

- Canada wild rye: hairless tiller; auricles wrap around tiller; short, ragged ligule; rolled emergent leaf

Round tiller near base, auricles absent, hair present on leaf collar

- Switchgrass: hairless tiller, tuft of hair inside leaf collar (may need hand lens), leaf midrib prominent, rolled emergent leaf

Tall dropseed: round and hairless tiller, coarse hairs in leaf sheath (may need hand lens), very short ligule with ciliate top, underside of leaf is smooth, top of leaf is somewhat rough to the touch, rolled emergent leaf

FORBS

Round stem, alternate leaves, hairlike stipules

Leadplant: circular simple leaves on young seedlings, odd-pinnate compound leaves on older seedlings, pointed leaflet tips

Purple prairie clover: crushed leaves have a citrus odor; 3 and 5 narrow, oblong leaflets

Showy tick trefoil: circular simple leaves on young seedlings, compound leaves with 3 leaflets on older seedlings, fine hairs on leaflet margins (hand lens)

Round-headed bush clover: compound leaves with 3 leaflets, pointed leaflet tips, hairy stem and leaf margins

Round stem, alternate leaves, leaflike stipules

Cream false indigo: compound leaves with 3 balloon-shaped leaflets; thick, fleshy leaves; hair on stem and leaves

Partridge pea: even-pinnate compound leaves, hairs on stem and leaves

White wild indigo: compound leaves with 3 balloon-shaped leaflets; thick, fleshy leaves; no stem or leaf hairs

Round stem, alternate leaves, stipules absent

New England aster: leaves clasp around stem, spatula-shaped leaves, soft hairs on stem and leaves

Round stem, opposite leaves, stipules absent

Butterfly milkweed: hairy stem; no milky sap; hairy, straplike leaves; leaf veins on cotyledon leaves

Edged stem, opposite leaves, stipules absent

Ox-eye sunflower: leaves with serrate margins; fine hairs cover leaves, making them rough to the touch

Wild bergamot: leaves with serrate margins smooth to the touch, mint odor of crushed leaves, underside of leaves often have purple coloration

Stem absent, leaves grasslike

Prairie blazing star: without petioles, multiple linear veins near leaf tip (hand lens)

Rattlesnake master: very visible individual hairs along leaf margins, petiole absent, fibrous leaves difficult to tear

Stem absent, leaves not grasslike

Compass plant: large, lance-shaped leaves; very large (pumpkin-size) cotyledons; very coarse hairs on leaves and petioles make them rough to the touch; prominent midvein on leaf

Golden alexanders: leaf shapes vary as seedling matures: first-year plants have apple-shaped (reniform) leaves with serrate margins and long petioles; narrow, straplike cotyledons with pointed tips

Gray-headed coneflower: coarse hairs on petioles and leaves make them rough to the touch; leaf margins can be entire, toothed, and deeply lobed; leaf blade base often not symmetrical

Pale purple coneflower: lance-shaped leaves with long petioles, leaves are very rough to the touch, 3 linear veins prominent on leaves

Stiff goldenrod: small notches on leaf margins (crenate), very fine hairs on leaf margins (hand lens) make them somewhat rough to the touch, leaves appear glossy

Equipment Needed for Field Sampling

Make sure you have the following equipment before you go into the field.

Seeding plan, which should include a species list, noting the quantities that were seeded; a description of the site preparation, seeding method, and sowing time; and an aerial photograph and soil map of the site.

Prairie seedling photographs and seed and seedling key such as *The Tallgrass Prairie Center Guide to Seed and Seedling Identification in the Upper Midwest* (Williams and Butler 2010).

Field guide to identifying prairie grasses and forbs for your region.

Weed identification guide such as *Weeds of Nebraska and the Great Plains*, published by the Nebraska Department of Agriculture.

Knee pads: Hard ground is tough on the knees.

Hand lens: Hairs, ligules, and auricles on seedling plants need magnification to be detected.

Digital camera and spare batteries.

Clipboard to supply a hard surface to record data.

Mechanical pencils: Mechanical pencils maintain sharp lead, which makes data recording easier.

Fig. 7-3. (*Facing page*) Grass morphology. Drawing by Brent Butler.

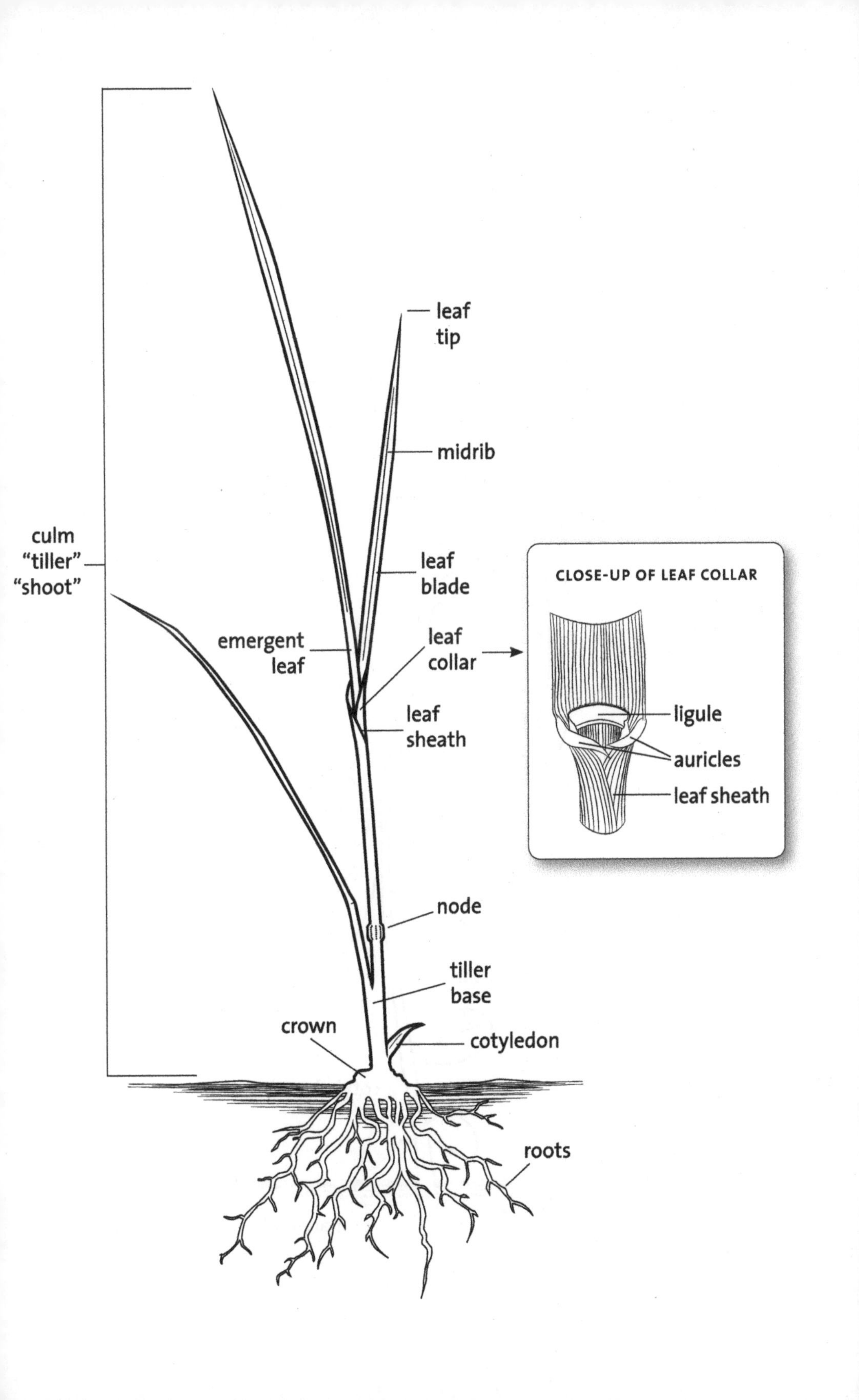
leaf
tip
midrib
culm
"tiller"
"shoot"
leaf
blade
CLOSE-UP OF LEAF COLLAR
emergent
leaf
leaf
collar
ligule
leaf
sheath
auricles
leaf sheath
node
tiller
base
crown
cotyledon
roots

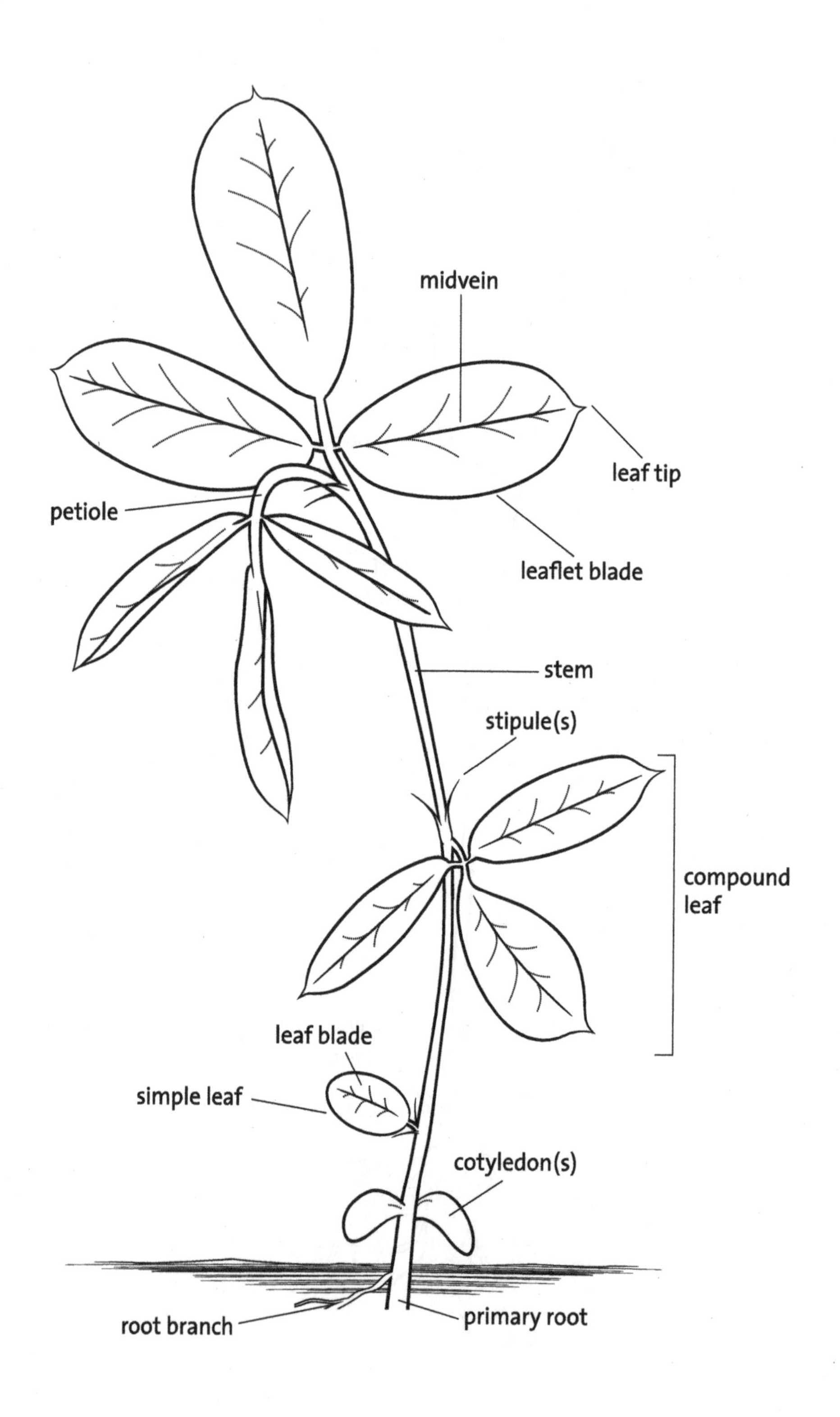
midvein
leaf tip
petiole
leaflet blade
stem
stipule(s)
compound
leaf
leaf blade
simple leaf
cotyledon(s)
root branch
primary root

Blank data entry sheets.
Quadrat frame.
Vinyl flagging for marking areas on the site that need reseeding, weed and brush control, etc.
Scissors to clip plant specimens. Fiskars are very durable.
Paper bags to hold plant specimens.
Fine-tipped Sharpie to mark bags for plant specimens.
Calculator.
Gloves prevent skin contact with plants that cause blistering like poison ivy and wild parsnip.
Insect repellent works well for mosquitoes and chiggers but not so well for biting flies.
Water so you can keep hydrated.
Sunglasses: The glare from the data sheets on a sunny day can be intense.

Scouting for Weeds

An additional tool for monitoring a new planting is to scout for weeds in the field. A good scouting effort will reveal weed patches not detected by the quadrat samples. It is critical to find weeds early before they increase in abundance and displace prairie plants. When scouting in the field, flag weed patches consisting of 20 or more individual plants. Mark their location and population size on the site map. This information is essential for the landowner to take the appropriate weed control measures.

Photographic Monitoring

A picture is worth a thousand words. In 1994, at the fourteenth North American Prairie Conference in Manhattan, Kansas, Paul Christiansen presented his research project of establishing prairie species in a roadside by overseeding them into nonnative smooth brome after a burn. He had established a permanent photo point on his research site and took images of the site before seeding and 2, 5, 10, and 13 years after seeding. Watching the plant community transform from a monoculture stand of nonnative grass to a diverse prairie plant community that resembled a prairie remnant was fascinating. Photographic monitoring can be extremely useful in reconstructed prairies to document the long-term

Fig. 7-4. (*Facing page*) Forb and shrub morphology. Drawing by Brent Butler.

vegetation changes. When establishing permanent photo points, consider the following.

Establish a permanent photo point for the camera mount. Use a sturdy metal fence post, and bury it at least 24 inches in the ground. The post should be 5 to 6 feet above ground. A metal post does not burn.

Establish a permanent reference point. Place a reference post in the site 50 to 100 feet away from the photo point post. Use a metal post installed as described above. Such a reference serves two purposes. First, it can be used as the focal point when the photographer takes the picture, to insure the same camera angle each time the site is photographed. Second, the reference point will assure the viewer that the photographs were taken at the same location.

Position the photo point post on the south side of the site and the reference point to the north. This will reduce sunlight glare.

Take the photograph on the same day and time each year or as close as possible, with the same light, several times during the season to show the changes in vegetation.

Include something for scale. A large tree or a fence line on the edge of the picture works well. A reference point could work.

Use a wide-angle lens.

Take a series of photographs during the growing season to detect different species (figs. 7-5 to 7-7).

Fig. 7-5. A 5-year-old prairie reconstruction in Dike, Iowa. Photo taken in late spring. Photo by David O'Shields.

Fig. 7-6. A 5-year-old prairie reconstruction in Dike, Iowa. Photo taken in mid-summer. Photo by David O'Shields.

Fig. 7-7. A 5-year-old prairie reconstruction in Dike, Iowa. Photo taken in late fall. Photo by David O'Shields.

Photograph the plants as they are flowering or just after they become dormant. Warm-season prairie grasses photograph best in late summer or early fall; cool-season grasses photograph best in early summer. Photograph woody plants after their leaves have dropped. This will allow you to count individual stems in the photograph and track their abundance.

Keep detailed records. Record when, where, and at what time the photograph was taken. Record any management activities at the site. Keep an ongoing species list, and note which species are abundant. Identify the species in weed and brush patches and label this on the photographs.

Summary

- Determining native plant establishment in year 1 or year 2 can save the landowner money by preventing unnecessary reseeding or allowing a targeted reseeding of smaller bare areas.
- Identifying prairie seedlings is essential for assessing prairie plant establishment (see chapter 16).
- A quadrat frame that has an area of 1 square foot inside the frame is often used to sample grassland vegetation.
- Plant density involves identifying and counting each prairie plant within the quadrat frame and is a reliable way to assess prairie plant establishment in a newly planted prairie.
- Plant frequency involves identifying each plant species within the quadrat frame; it is also a reliable way to assess prairie plant establishment in a newly planted prairie.
- A minimum of 20 to 30 quadrat samples will be needed to assess prairie plant establishment.
- Different physical characteristics of a site may cause varying prairie plant establishment.
- Areas with similar site characteristics should be sampled as a group and evaluated separately from other groups.
- The best time to sample a new planting is in late August to early September, when prairie seedlings will be more developed and easier to identify.
- Vegetation sampling can be done by one person. However, using two people, with one person recording data and the other person identifying seedlings, makes sampling easier.
- Order the prairie species on the data sheet first by group (grasses, forbs) and then by number of seeds sowed (highest to lowest).
- Record persistent perennial weed species.

- The presence of persistent perennial weeds in any quadrat sample should trigger immediate action to control these plants.
- A good scouting effort in the field will reveal persistent perennial weed patches usually not detected in the quadrat samples.
- Photographic monitoring can be extremely useful in reconstructed prairies to document the vegetative changes in nonnative and native plants over time.
- The quadrat sampling method discussed in this chapter is intended for the practitioner and may not be appropriate for research projects.

Prairie Restoration and Management

8

Identifying and Assessing Remnants

GREG HOUSEAL

Overview

Growing public awareness of tallgrass prairie will, we hope, lead to the identification, preservation, and management of more prairie remnants. Efficient use of limited resources requires strategies for locating, preserving, and managing these remnants. This chapter will highlight the value of prairie remnants, discuss where they might be found on the midwestern landscape, and consider attributes that affect remnant quality for preservation and long-term management.

What Is a Remnant?

Remnants are relict prairies or original pieces of the presettlement landscape. They have undisturbed or relatively undisturbed soil, though some soil disturbance may have occurred in the past. Undisturbed soil was never plowed, graded, or buried by fill. Other sites may have had brief soil disturbance in the past, for example, railroad beds that were graded in the 1800s or fields that were cultivated for brief periods and then abandoned. The key point regarding remnants is that some component of the original native vegetation remains. In the case of undisturbed soil, the original native vegetation may still be intact, or at least remnants of it have persisted on the site since settlement. At sites with limited soil disturbance, original native vegetation recolonized the site, either from residual rootstock and seed bank left on the site or reinvaded from surrounding intact prairie vegetation remaining at the time of disturbance. Original native vegetation is meant to imply that populations of species have persisted or regenerated themselves on the site through time (that is, they were not brought to the site and planted by people as in prairie reconstruction).

Irreplaceable Islands of Biodiversity

Remnants of tallgrass prairie are fragments, or islands, of biodiversity remaining after the large-scale destruction of the prairie ecosystem. As relicts of the original landscape, all remnants have biological, ecological, and cultural value and are deserving of preservation and management. Remnants may contain once-common animal and plant species that are now threatened with extinction. They may harbor rare species found nowhere else or populations of species with unique genetic traits and adaptations.

Remnants are also benchmarks to gauge the success of restoration efforts in terms of species composition and ecosystem functions. The untilled soils of remnants are the gold standards of fertility, soil structure, and soil biodiversity with which to compare agricultural soils. Ultimately, prairie reconstruction would not be possible without the seed sources and ecological information that remnant prairies offer. This, in part, is why identifying, assessing, and preserving remnant prairie is so vital. Beyond outright destruction, the greatest perils to the biodiversity of small remnants are continued isolation from gene flow and their vulnerability to either disturbance from surrounding land use activities or misguided management activities within the remnant. Buffering, reconnecting, and restoring prairie on a landscape scale is critical if native remnant tallgrass prairie is to be preserved as a viable ecosystem.

Locating Remnants

Where might remnants of the original prairie landscape persist in the highly fragmented and intensely farmed landscape of the modern Midwest? The process of ecosystem fragmentation is not random (Saunders et al. 1991). The social, economic, and political forces that fueled the conversion of tallgrass prairie to row-crop agriculture targeted the most accessible and productive areas first. Conversion of less accessible and less productive areas continued as technology (tiling, mechanized agriculture) and settlement allowed (Smith 1992). Aldo Leopold, in his day, observed that prairie plants were "content with any roadside, rocky knoll, or sandy hillside not needed for cow and plow" (Callicott and Freyfogle 1999). These remain likely places to look for remnant prairie.

For example, early transportation corridors (that is, rights-of-way) for roads and railroads are likely places for remnant prairie to persist. In the Midwest, all the original prairie landscape was fenced, and much of it was either grazed or farmed, after Euro-American settlement began in the mid 1830s. Yet prairie still existed along railroad corridors while railroads were being built during this same era. Sites where soil was graded with horse-drawn equipment to form the roadbed were recolonized by prairie from rootstock, rhizomes, and nearby seed sources. Both recolonized sites and intact remnants can still be found in pockets along railroads and are of equal value for preservation. Likewise, some of the road rights-of-way that were constructed earliest still harbor patches of remnant prairie.

Prairie species may persist in pastured areas, especially in hilly terrain or very wet areas that discouraged heavy grazing by domestic livestock. Sites with poor, thin, or very dry soil may harbor native species because they are grazed less intensely, since forage is less abundant. Prairie may persist in out-of-the-way corners of farm fields cut off by creeks or otherwise inaccessible to farm equipment and protected from herbicide drift (figs. 8-1 and 8-2).

Historic cemeteries are also likely places to find remnant prairie vegetation. They were often established on prominent hilltops and were fenced off from grazing livestock. Many of these sites have been mowed at times in the past but recovered when mowing ceased, or the prairie plants survived in the surrounding fence line. A few are preserved as prairie and maintained by volunteers and county or state resource managers.

Until the mid 1900s, some prairies were harvested for hay once a year in midsummer, since prairie hay was prized as high-quality forage for workhorses. Many of these prairie hay meadows were lost with the widespread mechanization of farming after World War II, when horses were no longer needed. A few examples of hayed prairie meadows remain in areas that were too wet, rocky, or small to row-crop, or in areas where the landowner preserved the practice as a cultural tradition. Some of these have since been preserved by county, state, and private nonprofit conservation organizations, but many remain in private ownership, often unrecognized as remnant communities.

Areas now covered with trees and brush may once have been open woodlands, savannas, or open prairie because fire and grazing kept woody growth in check. The presence of eastern red cedar trees may indicate thin or dry soil where prairie may persist. If not too densely shaded, openings and edges may yet harbor native prairie species. In the former oak/hickory savannas of the Ozarks in Missouri, trail corridors maintained for hiking allow sunlight into the understory immediately adjacent to the trail. These areas may be strewn with pale purple coneflower, prairie blazing star, and other prairie species, while none of these species will be found in the deep shade just a few feet off the trail (fig. 8-3).

Aerial photographs, particularly infrared, can help pinpoint likely areas to field-check for prairie remnants (fig. 8-2). Skill is required to interpret both the colors and the textures of these photographs. Knowledge of the vigor and

Fig. 8-1. Smith Prairie, a high-quality remnant prairie in Kossuth County, Iowa, is protected by a creek (just beyond the brushy area) that isolated it from farming activities. Photo by Greg Houseal.

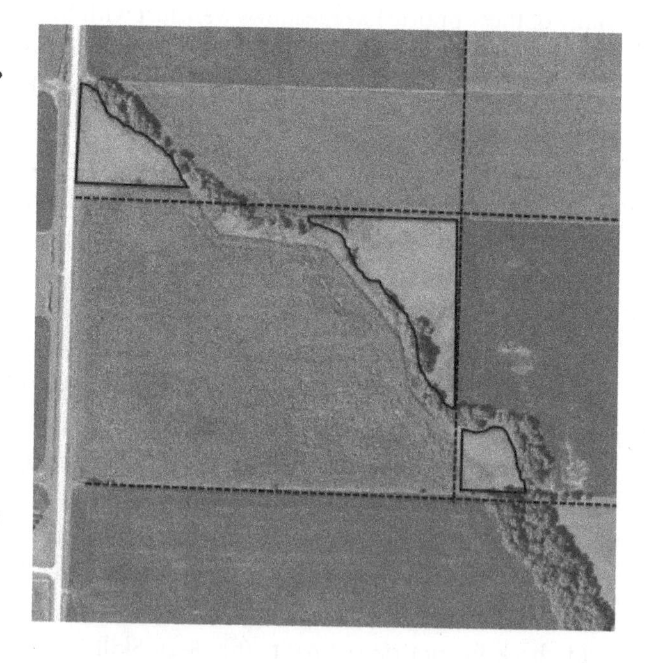

Fig. 8-2. Aerial photo of Smith Prairie, 2006, with conjectured historic property boundaries (dashed lines). Note how the creek effectively cut off equipment access for farming activities, preserving remnant areas (solid lines) for later county conservation board acquisition and restoration. Photo courtesy of Iowa State University Geographic Information Systems Support and Research Facility.

Fig. 8-3. Oak savanna recently cleared of brush with recovering remnant understory vegetation. An opportunistic weed, common mullein, blooms in the lower right-hand corner. Photo by Greg Houseal.

density of vegetation and the time of year of the photo is key to interpreting the colors of infrared aerial photography. Red is usually associated with live vegetation. Very intense reds indicate dense vegetation growing vigorously at the time the photo was taken. A cool-season pasture in spring or an irrigated alfalfa field in summer is an example of such vegetation. Dormant warm-season grasses (indicative of one type of prairie) in late fall or early spring may appear in shades of greens or tans. The pervasive presence of cool-season exotic grasses (for example, Kentucky bluegrass, smooth brome) may give a pinkish gray color to remnant prairies. Varying textures indicate the types of land cover (trees vs. grass vs. bare soil). In any case, it's critical to field-check potential sites. In Iowa, aerial photographs, including historic, black and white, and infrared, are available from the Iowa Geographic Map Server at http://cairo.gis.iastate.edu/. Aerial photographs are also available at local Natural Resources Conservation Service (NRCS) offices.

In summary, awareness of where remnants are likely to persist on the landscape, experience recognizing native plants, and aerial photo interpretation skills are all useful tools for locating remnants. Perhaps the most effective method,

Table 8-1. Indicators of the Quality of Remnant Prairies

Indicator	Low Quality	High Quality
Native species diversity	low	high
Presence of conservative species	absent	present
Soil profile	disturbed	undisturbed
Past site history	high impact	low impact
Invasive species	abundant	few/absent
Exotic species	abundant	few/absent
Aggressive woody species	dominant	minimal

however, is to seek out local knowledge from landowners, hunters, and native plant enthusiasts familiar with the area of interest.

Why Is Assessment Needed?

Once a remnant is located, assessment is needed to guide its management and rehabilitation; the assessment gives a measure of the quality of a remnant. Major factors that influence quality include native species diversity, particularly the presence of conservative species (those most sensitive to disturbance); prior management history of the site (grazing, overseeding, tiling, grading, etc.); and the presence of exotic or invasive species that pose an immediate threat to the remnant (table 8-1). There are three main objectives of remnant assessment: to determine appropriate management and rehabilitation strategies for the site (that is, do no harm); to monitor the recovery of the remnant in response to management activities; and to prioritize resources for acquisition, preservation, rehabilitation, and management of remnant sites.

NATIVE SPECIES DIVERSITY

The presence of original native vegetation is a key component of remnant quality. The more native species present, the greater the quality of the remnant. Ecologists refer to the number of species present on a site as species richness. A site with 80 plant species has greater species richness than a site with 20 plant species.

The types of species present and their abundance and distribution are also important considerations of quality. Species diversity, defined as the relative abundance of species throughout the site combined with the number of species present, gives a more complete description of remnant quality. Two prairies of the same native species richness (number of species) may differ considerably in diversity if one site has only a few individuals of many species, while the other has many well-represented individuals of each species.

Experience identifying native plants is critical to taking an inventory of remnant sites. Familiarity with only common species can lead to undervaluing a site containing rare species. Another difficulty is that native prairie spans a range of conditions, from wet to dry, and not all sites have the same species composition; every remnant is unique. There is a continuum of species along the hydrologic gradient such that a wet prairie looks nothing like a dry prairie. Likewise, many other types and subtypes of remnant native plant communities exist in the Upper Midwest, including wetlands (fens, bogs, seeps, sedge meadows, etc.) and woodlands (forests, open woodlands, savannas).

In Iowa, for example, the NRCS uses a checklist of approximately 25 to 30 native species for each of the four types of remnant prairie: wet, dry, mesic, and fen (a special case of a rare type of wetland) (table 8-2). These lists were developed for use on sites that are predominantly in a natural state and not highly disturbed; that is, some grazing may have occurred, but not enough to eliminate more conservative species. Not all species are required to be present, but each list includes some species that are more common and would be likely to persist in that remnant type. The checklist of native species combined with a site history checklist (table 8-3) is used by NRCS field technicians to quickly assess a site's potential as a remnant for the purpose of implementing USDA NRCS conservation practices. Promising sites are referred to trained ecologists and botanists for further investigation.

If natural areas are to be compared, inventories of consistent scope and precision must be conducted. A thorough plant inventory requires at least monthly surveys throughout the growing season. Factors that will affect the total number of species identified include the skills of the observers, the number of observers, and the amount of time spent surveying the site. It is important to apply equal effort toward each inventory so that meaningful comparisons can be made between sites.

Thoughtful and informed management of the site and repeated visits over subsequent years will increase the odds of observing additional species. Certainly any federal- and state-listed threatened and endangered species should be noted. Federally listed species are protected by law, and their presence may help secure funding for acquisition or management of the site. Lists of federal and state threatened and endangered species are available from the U.S. Fish and Wildlife Service at http://www.fws.gov/Endangered/wildlife.html.

Table 8-2. NRCS Remnant Species Checklists for Assessing Remnant Potential

Remnant Species Checklist for Wet Sites

Scientific Name	Common Name
Agalinis tenuifolia	Slender False Foxglove
Amorpha fruticosa	Indigo Bush
Anemone canadensis	Canada Anemone
Asclepias incarnata	Swamp Milkweed
Calamagrostis canadensis	Bluejoint
Carex stricta	Tussock Sedge
Cephalanthus occidentalis	Buttonbush
Eupatorium maculatum or Eupatorium perfoliatum	Spotted Joe-pye-weed or Common Boneset
Glyceria striata	Fowl Mannagrass
Helenium autumnale	Sneezeweed
Impatiens capensis	Spotted Jewelweed
Iris virginica var. shrevei	Blue Flag Iris
Lobelia cardinalis or L. siphilitica	Cardinal Flower or Great Blue Lobelia
Lysimachia ciliata	Fringed Loosestrife
Lythrum alatum	Winged Loosestrife
Mentha arvensis	Common Mint
Mimulus ringens	Monkey Flower
Oligoneuron riddellii	Riddells' Goldenrod
Onoclea sensibilis	Sensitive Fern
Phyla lanceolata	Fogfruit
Sparganium eurycarpum	Common Bur Reed
Spartina pectinata	Prairie Cordgrass
Stachys pilosa	Hedgenettle
Thalictrum dasycarpum	Purple Meadowrue
Thelypteris palustris	Marsh Fern

Remnant Species Checklist for Mesic Sites

Scientific Name	Common Name
Allium canadense	Wild onion
Andropogon gerardii	Big Bluestem
Astragalus canadensis	Canada Milkvetch
Baptisia lactea	White wild indigo
Ceanothus americanus	New Jersey Tea
Coreopsis palmata	Prairie Tickseed
Desmodium canadense	Showy Tick Trefoil
Eryngium yuccifolium	Rattlesnake Master
Heliopsis helianthoides	Ox-eye
Hypoxis hirsuta	Yellow Star Grass
Liatris pycnostachya	Prairie Blazingstar
Lithospermum canescens	Hoarry Puccoon
Lobelia spicata	Palespike Lobelia
Penstemon digitalis	Beardstongue
Phlox pilosa	Prairie Phlox
Potentilla arguta	Prairie Cinquefoil
Pycnanthemum tenuifolium	Slender Mountain Mint
Pycnanthemum virginianum	Mountain Mint
Ratibida pinnata	Gray-headed Coneflower
Silphium laciniatum	Compass Plant
Sorghastrum nutans	Indian Grass
Teucrium canadense	American Germander
Tradescantia ohiensis	Spiderwort
Tripsacum dactyloides	Eastern Gamagrass (Southern Iowa)
Veronicastrum virginicum	Culver's Root

Remnant Species Checklist for Xeric Sites

Scientific Name	Common Name
Allium canadense	Wild Onion
Amorpha canescens	Leadplant
Anmeone cylindrica	Windflower
Asclepias tuberosa	Butterfly Milkweed
Baptisia bracteata	Long-bract Wild Indigo
Bouteloua curtipendula or Bouteloua gracilis	Side-oats Grama or Blue Grama
Brickellia eupatorioides	False boneset
Ceanothus herbacea	Redroot
Coreopsis palmata	Prairie Coreopsis
Dalea candida or D. purpurea	White Prairie Clover or Purple Prairie Clover
Echinacea pallida or E. purpurea	Pale Purple Coneflower or Purple Coneflower (Purple coneflower restricted to southern Iowa)
Euphorbia corollata	Flowering Spurge
Hesperostipa spartea	Porcupine Grass
Koeleria macrantha	June Grass
Lespedeza capitata	Round-headed Lespedeza
Liatris aspera, L. punctata, or L. squarrosa	Rough Blazing Star, Dotted Blazing Star, Scaly Blazing Star (Dotted Blazing star in western Iowa)
Lithospermum canescens, L. caroliniense, or L. incisum	Hoary Puccoon, Carolina Puccoon, or Fringed Puccoon
Oligoneuron rigidum	Rigid Goldenrod
Pedicularis canadensis	Wood Betony
Schizachyrium scoparium	Little Bluestem
Silphium integrifolium	Rosinweed
Sisyrinchium campestre	Stout Blue-eyed Grass
Sporobolus clandestinus or S. heterolepis	Rough Dropseed or Prairie Dropseed
Symphyotrichum oolentangiense	Skyblue Aster
Viola pedata or V. pedatifida	Birds-foot Violet or Prairie Violet

Remnant Species Checklist for Fen Sites

Scientific Name	Common Name
Agalinus purpurea*	False foxglove
Calamagrostis canadensis	Canada Bluejoint
Caltha palustris	Yellow Marsh Marigold
Carex hystericina or C. stricta	Bottlebrush Sedge or Tussock Sedge
Chelone glabra	White Turtlehead
Cicuta maculata*	Water hemlock
Doellingeria umbellata*	Flat top aster
Eriophorum angustifolium	Tall Cottongrass
Eupatorium perfoliatum or E. maculatum	Common Boneset or Spotted Joe-pye-weed
Gentianopsis procera*	Small fringed gentian
Geum aleppicum	Yellow Avens
Helenium autumnale	Sneezeweed
Lobelia kalmii* or L. siphilitica	Kalm's lobelia or Great Blue Lobelia
Lycopus americanus or L. uniflorus	American Water Horehound or Northern Bugleweed
Lysimachia quadriflora	Four Flower Yellow Loosestrife
Muhlenbergia glomerata*	Marsh wild timothy
Oligoneuron riddellii*	Riddell's goldenrod
Onoclea sensibilis	Sensative Fern
Parnassia glauca*	Grass of Parnassus
Pedicularis lanceolata	Swamp Lousewort
Pycnanthemum virginianum	Virginia Mountain Mint
Schoenoplectus pungens*	Three-square
Spiraea alba	White Meadowsweet
Thelypteris palustris	Eastern Marsh Fern
Triglochin maritimum*	Common arrow-grass

* Potential fen indicator species

Note: Species listed are considered common and are frequently encountered in remnants. No species with a coefficient of conservatism below 3 was included, to exclude colonizing by native weed species that may occur in degraded or highly altered conditions; although native, these weed species — such as common evening primrose, hoary verbena, and Canada goldenrod — do not indicate a remnant community.

Source: Table provided by Iowa NRCS staff biologist Jennifer Anderson-Cruz.

PRIOR LAND USE HISTORY

It is important to obtain any available information regarding prior land use at a site (fig. 8-4). As our knowledge and ability to reconstruct prairie improve, distinguishing low- or moderate-quality remnants from high-quality reconstructions will become more difficult. While this is something to celebrate, a planting is still only a reproduction of the original prairie. A degraded remnant with few and scattered native species has more intrinsic value than a high-quality, species-rich planting and should be preserved and restored. This is in part due to the likelihood that a remnant may harbor previously undocumented species —plant or animal—and be a repository for populations with potentially unique genetic adaptations. Important impacts to consider and evaluate are the history and extent of cultivation, grazing, herbicide use, changes in the hydrology (tiling, etc.), and interseeding of both native and nonnative species.

The landowner, local residents, and conservation organizations should be consulted for information on past land use of a site. Historic aerial or landscape photographs should be sought and scrutinized for the valuable information they may contain. Original land survey records from the 1800s can shed light on whether an area was considered a prairie, savanna, woodland, or wetland at the time of the survey. This information can be used to guide rehabilitation efforts. (Iowa land records are available at http://www.public.iastate.edu/~fridolph/dnrglo.html.)

EXOTIC OR INVASIVE SPECIES

The presence of exotic (nonnative) or invasive species on a remnant site will need to be factored into the potential cost of and strategies needed for rehabilitation and long-term management of a remnant. Data from the Illinois Natural History Survey indicate that nearly 30 percent of the state's current flora is introduced (nonnative) (Robertson et al. 1997). Robertson et al. (1997) grouped introduced weed species into three general categories (table 8-3):

1. Invasive: extremely serious weeds that out-compete native species and threaten to destroy natural plant communities in prairies
2. Persistent: species that occur regularly but are not overly aggressive and are not likely to significantly change species composition in prairies
3. Opportunistic: species that could probably be eliminated with proper management practices

Weed species in the first category, invasive, will lower the remnant's quality over time and can present significant challenges to long-term restoration and

Fig. 8-4. NRCS Site History Checklist for Iowa

Remnant and Fen Site History Checklist

____ 1. Remnant/fen boundaries delineated on an aerial photo (CIR-Ortho map via GIS preferred).

____ 2. Provide remnant acreage on hydric soils, acreage on non-hydric soils, and acreage on inclusions/complex soils.

____ 3. If an inventory of the site's vegetation has been conducted, provide a species list indicating the number of visits, dates, and an approximate number of hours spent compiling the list. If such a list is not available, indicate on appropriate remnant species checklist which of the listed species occurs within the site.

____ 4. Has the site been recognized as unique by a non-government organization, local, county, state or federal entity? Explain.

____ 5. History and extent of cultivation.

____ 6. Past and current hydrologic manipulation.

____ 7. Grazing history and intensity.

____ 8. Herbicide usage.

____ 9. Has the site been inter-seeded, if so, when and with what?

____ 10. Extent of coverage by exotic/invasive species (cool season grasses, forbs, noxious weeds, etc.).

____ 11. List any unique or rare animal species found within the site or in immediately adjacent areas.

____ 12. List potential or known external threats to the site:
a. non local ecotype native seeding within ½ mile
b. lack of adequate buffer (<200 ft)
c. other

____ 13. GPS location and acreage of the prairie remnant of fen (optional, required upon acceptance).

____ 14. If a fen, indicate the depth of the *organic* soil horizon.

Table 8-3. Common Weed Species of Tallgrass Prairie

Invasive Species	Persistent Species	Opportunistic Species
Canada thistle	smooth brome	chicory
Crown vetch	musk thistle	bull thistle
Queen Anne's lace	ox-eye daisy	smooth sumac
Cut-leaved teasel	tall fescue	common mullein
Common teasel	white sweet clover	common ragweed
Leafy spurge	yellow sweet clover	great ragweed
Sericea lespedeza	wild parsnip	
Purple loosestrife	Kentucky bluegrass	
Pampas grass	multiflora rose	
Reed canary grass	poison ivy	
Common buckthorn	red clover	

Source: Modified from Robertson et al. 1997.

management of the site. Applying no management to the site means losing the remnant plant community to the invasive species, yet control methods used on invasive species may in themselves be detrimental to the remnant. Several resources are available with regional information on invasive species, their biology, and control methods; see, for example, the USDA National Agricultural Library, http://www.invasivespeciesinfo.gov, and the National Park Service, "Weeds Gone Wild," http://www.nps.gov/plants/alien/.

Weed species that are persistent and opportunistic are less of a problem but still require management. Reed (2004) suggests mapping areas with similar plant community types (patches) within the remnant and surrounding adjacent areas. Consider the species diversity of each patch and its potential risk for invasion by weed species on or near the site. Also note edges and corridors that are or may become entry points for invasive species, and develop strategies to combat them in these locations. Such a map will facilitate setting specific and measurable restoration goals for a remnant site. High-quality remnant areas or areas containing endangered species most susceptible to invasion should be given higher priority for management.

Other Factors Affecting Remnant Quality and Management

Other factors that may affect the remnant's quality and management include the size and shape of the remnant, distance and connectivity to other remnants, and land use surrounding the remnant (Saunders et al. 1991). The smaller the remnant, the greater the impact that external forces (invasive species, herbicide drift, nutrient and water influx) will have on its quality and long-term survival. Larger remnants are likely to have greater diversity because they are more likely to encompass different types of habitat, yet high-quality remnants as small as 10 acres may possess most of the local diversity present in a much larger prairie (Robertson et al. 1997). The size of a remnant also determines the potential population size of a species. Larger populations tend to have greater levels of genetic diversity and thus may be more resilient (adaptive) when faced with environmental stressors and more resistant to extinction (Gilpin and Soule 1986). There is also evidence that seed viability increases with larger populations, possibly because they attract more pollinators and/or are more genetically diverse (Menges 1991).

The shape of a remnant has more significance for smaller rather than larger remnants, mainly because size affects the perimeter–to–core area ratio of a remnant (Diamond 1975). Long, thin strips of vegetation may be associated with a natural feature (river, creek corridors) or a consequence of transportation or utility corridors. Linear remnants may contain greater species diversity than block-shaped remnants if they traverse different topographic features and therefore contain different plant communities (Saunders 1991). However, the large perimeter-to-area ratio of a linear remnant will likely present greater management challenges and a higher cost per unit area to mitigate negative impacts from adjacent land use activities. Opportunities for reconstructing or restoring buffer areas around the core remnant should be considered.

The distance from other remnants affects a remnant's ability to recruit new species. Gene flow from one remnant to another via pollen and seed transport may be cut off with greater distance, leading to declining genetic diversity within the remnant populations. Opportunities for restoring or reconstructing prairie in the intervening areas should be considered, since corridors of similar plant communities connecting remnants aid the migration of animal and plant genes (via transport of seed and pollen) from one remnant to another. Corridors also enhance the quality of remnant areas by providing extra foraging areas for wildlife and serve as refugia for plant and animal species during disturbance, such as prescribed fire or other management activities.

Potential negative impacts upon remnants from surrounding land use should be considered, as these will greatly affect future remnant quality and exacerbate management challenges. For example, low-lying remnants are water run-in sites (contrasted to upland run-off sites) and will be affected by the water quality (nutrients, herbicide and pesticide residue) and sediment (including seed and plant propagules) influx from upslope land use activities. Many otherwise high-quality prairie remnants have had swales overrun with sediment and associated invasive species (such as reed canary grass from farming activity upslope). Mitigating upslope disturbances with buffer strips of native vegetation will help preserve remnant quality.

Loss of habitat coupled with the dominance of crop species on the surrounding landscape may concentrate herbivores and pest species in small remnants. Smaller herbivores generally are more selective grazers than larger herbivores and can decrease plant diversity in remnant areas (Miller et al. 1992; Howe 1999). Whitetail deer, for instance, will selectively graze flower stalks of white wild indigo and swamp saxifrage, precluding seed production. Voles are more common in cool-season pastures and hayfields and can negatively affect plant species in adjacent prairie remnants and reconstructions (Lindroth et al. 1984; Howe and Brown 1999). Jameson (1947) reports finding a vole's winter cache of up to 6 quarts of corms of two *Liatris* species in an Illinois prairie. Mitigating these negative impacts on small isolated remnants by modifying surrounding land use will enhance the quality of the remnant areas being preserved.

Summary

- The efficient use of limited resources requires strategies for locating, preserving, and managing remnant prairies.
- Remnants or relict prairies have relatively undisturbed soil and at least some component of the original native vegetation.
- Original native vegetation is meant to imply that populations of species have persisted or regenerated themselves on the site through time.
- Remnants of tallgrass prairie vegetation are fragments, or islands, of biodiversity remaining after the large-scale destruction of the ecosystem.
- As relicts of the original landscape, all remnants have biological, ecological, and cultural value and are deserving of preservation and management.
- Remnants are benchmarks to gauge the success of restoration efforts in terms of species composition and ecosystem functions.
- Remnant prairie may persist in old road or railroad transportation corridors, lightly grazed pastures, out-of-the-way corners of farm fields, prairie hay meadows, pioneer cemeteries, and openings and edges of brushy areas.
- The three main objectives of assessing remnant quality are to determine appropriate management and rehabilitation strategies for the site; to monitor the recovery of the remnant in response to management activities; and to prioritize resources for acquisition, preservation, rehabilitation, and management of remnant sites.

- Major factors that influence quality include native species diversity, species composition, prior management history of the site, and the presence of exotic or invasive species that pose an immediate threat to the remnant.
- The presence of original native species is a key component of remnant quality. The types and number of species present and their abundance and distribution are important measures of remnant quality.
- Knowledge and documentation of prior land use history of a site are important to distinguish a remnant from a reconstructed prairie.
- Prior land use impacts to consider and evaluate are the history and extent of cultivation, herbicide use, changes in the hydrology (tiling, etc.), and interseeding.
- A degraded remnant with few and scattered native species has more intrinsic value than a high-quality, species-rich planting and should be preserved and restored.
- Invasive species on a remnant will lower its quality over time and can present significant challenges to long-term restoration and management of the site.
- Additional factors affecting quality include the size and shape of the remnant, distance and connectivity to other remnants, and land use surrounding the remnant.

9 The Restoration of Degraded Prairie Remnants

DARYL SMITH

Overview

Heavily grazed or brushy sites with relict prairie plants often appear to be so badly degraded or damaged that there is little hope of restoring the prairie. However, it is possible to restore these sites by eliminating the causes of degradation, repairing the damages, and adapting techniques used in prairie management. Depending on the condition of the degraded remnant, restoration measures may include extensive removal of invasive woody species, modification of the hydrology of the site, expansion of relict populations, reintroduction of species, elimination of invasive exotics, or a combination of these approaches. Proceed carefully, using a conservative approach. Opinions vary concerning whether extirpated species should be reintroduced. When the restoration goal for the site is attained, prairie management techniques can be used to maintain the prairie.

Planning

Improving an existing degraded prairie is one of the most important types of prairie restoration. The intent is to restore a site retaining relict prairie plants to a condition as close as possible to its original species composition and structure.

Degraded prairie remnants that have been heavily disturbed, modified, or damaged are candidates for this form of restoration. Evidence of degradation includes low species diversity and richness, lack of conservative species, and numerous invasive exotic weeds and/or aggressive woody species. These are the result of overgrazing, cultivation, suppression of fire, indiscriminate herbicide use, sedimentation, changes in hydrology, and interseeding of nonnative forage species. Such remnants are not recovering naturally by secondary succession and retain only rudiments of a native prairie. Examples include prairie remnants so overgrown with woody species that they appear to be impenetrable

Fig. 9-1. Woody species encroaching upon a prairie opening. Photo by Daryl Smith.

brush patches with a few relict prairie species in small openings, heavily grazed pastures with a few scattered prairie plants in which most of the native species have been replaced by nonnative, cool-season grasses and weeds, and prairies that have partially recovered from limited cultivation in the past. If no restoration measures are applied, these remnants will continue to degrade, and the prairie will eventually disappear (fig. 9-1).

High-quality prairie remnants with little or no degradation or those that have significantly recovered from past degradation through secondary succession need not be considered for special restoration. The stability and composition of these remnants can be maintained with proper prairie management.

A restoration is completed when the site is restored to the condition specified by the goals of the plan. Often this condition approximates the quality of the reference remnant. Thereafter, more routine prairie management techniques such as prescribed fire, spot spraying of herbicides, grazing, and mowing can be used to maintain the quality of the restored prairie.

Most individuals are anxious to begin restoring a remnant site soon after acquisition. Don't act hastily—mistakes made early in the restoration process can be difficult to correct. Restoration activities should not be initiated until you have a plan in place.

The following information is needed to develop a restoration plan. Much of this information will be available from the assessment of the degraded site (see chapter 8).

Determine the degree of degradation or damage to the site and its effect on the plant communities.

Prepare a comprehensive species list.

Identify rare and uncommon native species as well as problem species like invasive exotics and aggressive woody natives (for example, gray dogwood and smooth sumac).

Discuss the structure and composition of communities within the remnant.

Obtain information on soil, hydrology, and other aspects of the physical environment as well as presettlement vegetation from original land survey records and soil surveys.

Prepare a land use history; that is, gather information about cultivation, tiling, grazing, and soil disturbance from construction.

Selecting a reference remnant for the restoration is an important part of the planning process. The reference remnant serves as a model for the restoration and as a standard for monitoring and evaluating the project. A reference remnant should resemble the project site as closely as possible in terms of location, soil, topography, and hydrology. It is usually an existing remnant, although it could be a detailed ecological description of a former remnant.

Formulate project goals to provide the framework for implementing, completing, and monitoring the restoration. Goal setting should include a realistic assessment of the degree to which the restoration can replicate the desired original community or place it on a trajectory toward that original condition. Other important considerations in goal development are the extent of remnant degradation, knowledge of the former condition of the site, availability of plant materials, and funding capability.

Often, the overarching goal for a restoration plan is to return the degraded remnant to a predetermined, presettlement condition. However, it may be necessary to consider a more limited goal for many remnants. Exotic, cool-season species such as smooth brome and/or Kentucky bluegrass may be so pervasive on the site that they cannot be eliminated in the restoration process. Furthermore, many of the remnants on today's landscape are too fragmented and small to allow the restoration of a naturally functioning, self-regulating prairie ecosystem that is integrated with its surroundings. The only places where self-regulating restorations may be possible are certain sites within major landforms where large areas of prairie persist, such as the Flint Hills of eastern Kansas, the Osage Hills of north-central Oklahoma, the Loess Hills of western Iowa, and the glacial moraines of northeastern South Dakota and southwestern Minnesota.

A restoration plan should include the following minimal components recommended by *The SER International Primer on Ecological Restoration* (2004): rationale of need for restoration; ecological description of the site designated for restoration; statement of the goals and objectives of the restoration project; designation and description of the reference site; explanation of how the proposed restoration will integrate with the landscape; explicit plans, schedules, tasks, and budgets for initial site recovery, project development, and completion goals, including a strategy for making prompt midcourse corrections; monitoring protocols for evaluating the project; and strategies for long-term protection and maintenance of the restored ecosystem.

Restoration processes often vary from remnant to remnant or even within a single remnant. In some cases, relict vegetation of smaller remnants may be sufficiently uniform so that the same plan and techniques can be used throughout. However, differences in topography, hydrology, and soil on larger remnants can result in a variety of plant communities. The communities are often in different stages of degradation and require a plan that incorporates a variety of techniques for recovery.

Implementing the Restoration Plan

The following sequence of activities is typical of most restorations.

Halt the degradation of the relict prairie community. One of the first steps in the restoration process is to stop whatever is causing the degradation. For example, livestock must be removed from overgrazed sites to allow vegetation to recover. Woody species must be removed from overgrown woody thickets to allow light to reach shaded prairie vegetation. This is necessary to prevent further loss of prairie vegetation and, hopefully, begin recovery.

Repair physical site conditions that have been damaged or greatly modified. The physical conditions of some portions of the site may have been changed so much that there is no support for the original plant communities. For example, it may be necessary to restore the hydrology of the site—remove or break the field tile—to recreate the original moisture conditions. In cases of extreme soil disturbance, it may be necessary to restore the original soil profile by removing accumulated sediment or replacing eroded soil.

Retain legislatively protected species and species of special concern. Rare or uncommon species may require special protection, consideration, or care during the restoration process. Retention of populations of federal or state threatened or endangered species is critical to preserving the quality of a natural area.

Use prescribed fire to enhance native species and control some nonnative species. Prescribed fire can be used as soon as the vegetation develops a fuel load sufficient to carry fire. Often, two growing seasons are needed for vegetation recovery. Subdivide the site into burn units, and prioritize the areas to be burned. Low-quality areas with limited species diversity are good candidates for initial burns. Properly timed burns provide a competitive edge for prairie vegetation to displace weeds and increase the fuel load to support future prescribed burns.

Control invasive perennial weeds without harming existing natives. Annual weeds can be controlled with mowing to prevent seed production, but invasive perennials will require more aggressive treatment with appropriate herbicides. Targeted use of herbicides by spot spraying or wicking may be necessary to avoid harming native species. Hand-pulling weeds may accomplish the same purpose with smaller infestations (see chapter 6).

Increase the populations of existing native species. Seed can be collected from abundant relict species and spread throughout the communities that they inhabit. Small populations of relict species with low genetic diversity may benefit from the addition of seed from the reference remnant or other nearby sources.

Reintroduce species that have been extirpated. If replication of the composition of the original community is a goal, it may be necessary to reintroduce species that have been extirpated as a result of the degradation of the site. When reintroducing missing species, collect seed from nearby remnants of similar topography, soil, and hydrology.

Monitor the project. A monitoring plan should be in place from the onset of the restoration process. Establish permanent sampling transects and photographic locations to assess changes during the restoration process. Monitoring will provide feedback to make corrections during the restoration process and provide assurance that the project is progressing as planned. The information collected can also be used to evaluate the completed project. Monitoring is often overlooked in the restoration process. Include it in the plan so that you don't regret the lack of information later. Chapter 7 discusses monitoring techniques.

Control restoration costs. Estimate costs in advance and develop a budget to determine whether funds are available for completing the project. Highly degraded, low-quality remnants require intensive management that increases project expenses. Extension agencies should be able to provide custom cost rates to allow you to develop a budget for equipment rental and management activities. Grants to defray restoration costs may be available from state or local conservation organizations. Volunteer labor may be used to reduce costs but requires additional staffing and resources for coordination. However, volunteers can also be strong advocates for the project.

Identify and formulate strategies to insure long-term protection and management of the site. Most remnants are vulnerable to external degradation processes since they are surrounded by croplands dominated by nonnative vegetation. Adjacent areas upslope from a remnant can be troublesome sources of sediment containing invasive species and/or herbicide residue. Properly located buffers can effectively reduce invasion of exotic species and negative impacts from adjacent land uses. Retain your restoration investment by providing reasonable assurance that the restored remnant will be protected and properly managed in the future. Explore the possibility of using a conservation easement, as this is an effective means of insuring long-term protection of a site.

Reintroducing Extirpated Species

If the primary goal is to restore the presettlement species composition of a site, obviously it will be necessary to reintroduce species that have been extirpated. A decision to reintroduce a species should include some assurance that it was part of the original vegetation. Add only species known to have been extirpated from the site or those that were quite likely part of the original flora. A record of a species from the restoration site is the best assurance of its prior presence. If such a record is not available, the presence of the species in the reference site is a good indication that it was once part of the original community. Hold off reintroducing extirpated species until restoration is well under way, after degradation has been halted, prescribed fire has been initiated, and populations of relict species are recovering.

If the restoration site contains a high percentage of original conservative species, proceed cautiously by reintroducing a few species at a time to a limited part of the site. Usually, small areas within a restoration site are interseeded by hand or supplemented with seedling transplants; however, a native-seed drill is more effective for interseeding if the entire site contains a limited number of relict species. Maintain good records of reintroduced species so their origin is known when they appear. A good approach to adding a species is to interseed in the fall with seed collected from a nearby reference site with similar soil and topography. Similarity of the collection site to the restoration site is more important than the distance that separates them. If similar sites for seed collection are not nearby, obtain source-identified seed for your region from a native-seed grower. Growers of source-identified seed are required to maintain a record of the collection sites of their seed, and you can obtain a description of those sites.

The presence of a few individuals of conservative species on a site might delay a decision to add extirpated species. The concern would be that the added native species might be more aggressive and displace the rare conservative species. The restoration goal would be to improve the site to maintain and increase population numbers of the conservative species before considering the addition of extirpated species.

Opinions differ about reinroducing extirpated species to a site. Some ecologists maintain that introducing new genetic stock compromises the historic integrity of the relict plant community. Whether to reintroduce extirpated species may depend on the condition of the remnant, native species composition, and personal preference of the project manager. If only 3 or 4 common native species remain in a former heavily grazed pasture after sufficient recovery time, the decision to reintroduce species is relatively easy. However, there are no criteria relating minimum numbers of relict native species to the addition of extirpated species. You could decide that you wanted to maintain and manage a former pasture containing 15 to 20 native species without adding more species. You would stop the degradation, remove undesirable weeds, and institute management practices to enhance the existing native species. The goal would be to maintain a limited suite of prairie species rather than approximate the original composition of the remnant. After working with the existing vegetation for a period of time, you could then decide whether to add species.

Restoration Activities on Degraded Remnants

The first task in restoring overgrown, brushy sites with a few pockets of prairie species is to remove invasive trees and shrubs. This will open up the area to more light and allow prairie species to increase. Although summer burns are difficult to ignite, they can be quite effective in controlling woody species. Control techniques for woody species are discussed in chapter 10. Weed species are likely to initially occupy the site after removal of the tree and shrub canopy, as native species will not be sufficient to immediately fill the space. Annual weeds can be mowed to prevent seed production; however, herbicides may be needed to control invasive perennials. Initiate prescribed burning as soon as vegetation growth is sufficient to carry fire. Properly timed prescribed fire will stimulate the natives and curtail weeds (see chapter 10). Collect seed from prairie species on the site, and seed directly into the ashes soon after the fire. Begin assessing the plant community in the next growing season after the burn. Ongoing brush clearing and increased frequency of prescribed fire will enlarge openings and favor the expansion of prairie species. These more intensive restoration activities will help put the site on a trajectory toward the original prairie composition and structure.

Remove herbivores from heavily grazed sites to provide an opportunity for native vegetation to recover. It is difficult to determine the presence and abundance of native species until they have had sufficient time to grow, flower, and set seed. Most sites contain variations in topography and hydrology. Relict prairie species are more likely to persist in areas less frequented by grazers — steep slopes, marshy areas, and locations far from barns. Allow ample time for the appearance of species that were not observed in the initial assessment, and continue to survey the site for native species each growing season. Conservative species such as western prairie fringed orchid, Michigan lily, and prairie lily have reappeared after being absent from a site for many years. Properly timed prescribed fire will enhance existing prairie by stimulating the growth of native species (see chapter 10). A prescribed burn regime can be initiated as soon as there is sufficient fuel to support a fire. Initially, burn areas with few native species to accelerate their recovery. Control invasive perennial weeds by spot or wick application of herbicides to avoid killing existing natives. Hand-removing some weeds may be the only option even though this is time-consuming and labor-intensive.

Previously cropped remnants can usually recover to a large degree from the effects of cultivation. Most exist because they had been cultivated only once or twice and recovered from rootstock and/or species that persisted in the seed bank, or they were able to recover by secondary succession because prairie vegetation was growing nearby when cultivation ceased. The second type of recovery is similar to that which occurred in rights-of-way during railroad or road construction through existing tallgrass prairie. Such remnants often contain an eclectic collection of prairie plants with a higher percentage of deep-rooted species. Often, certain native species are missing, and those that persist are scattered in a matrix of nonnative species such as smooth brome. Therefore, locating a reference remnant to serve as a guide for species composition of the restoration is very important. A thorough baseline inventory of the restoration site is essential to guide decisions about which species may need to be reintroduced to restore the original plant community. Restoring formerly cultivated sites requires selective reduction or elimination of the matrix of nonnative species, increase of relict species, and creation of conditions favorable for the reintroduction of extirpated species. For example, if the nonnative matrix is an extensive stand of smooth brome, glyphosate may be applied when the cool-season brome is actively growing and prairie species are dormant. Frequent use of prescribed fire in late spring will aid in brome reduction.

Wildin Heritage Prairie: A Case Study

The story of the acquisition and initial restoration of Wildin Heritage Prairie in Kossuth County by the Iowa Natural Heritage Foundation (INHF) is a good example of prairie restoration as discussed in this chapter. The questions and concerns raised, information sought, techniques available, methods used, processes involved, and results obtained are representative of what would be encountered in restoring a degraded prairie in the Upper Midwest. Information for the story was provided by "Return of the Prairie," an article by Cathy Engstrom in the fall 2004 issue of the *Iowa Natural Heritage* magazine, and personal communication with Joe McGovern, land stewardship director for the Iowa Natural Heritage Foundation.

During the fall of 2002, the INHF board of directors and staff were trying to determine whether they should purchase a potential prairie site located on a glaciated portion of the Des Moines Lobe. The level site sprinkled with potholes contained some relict prairie species, had never been plowed or drained, and had only occasionally been spot-sprayed for Canada thistle. However, it had been heavily grazed and looked like a typical weedy, overgrazed pasture, causing the staff and directors to question whether it was a good candidate for purchase and restoration. In December 2002, after much deliberation, the Iowa Natural Heritage Foundation approved the purchase of the 80-acre pasture from the estate of George Wildin. "It was a leap of faith for both our board and staff," said Joe McGovern, "we saw a potential prairie remnant there, but we weren't positive we could bring it back."

Reservations of INHF staff and board members regarding the potential of the site for prairie recovery were well founded. The area had been heavily grazed annually for the past 12 to 15 years. The cumulative effect of grazing by cattle from May through early September followed by approximately 2 months of grazing by horses had taken its toll on native species. McGovern commented that prior to removal of cattle and horses in November 2002, even well-trained botanists would have had difficulty distinguishing prairie plants from non-native species among the closely cropped vegetation (figs. 9-2 and 9-3).

The most positive evidence of original prairie vegetation was found near a dozen or more undrained potholes scattered throughout the site. McGovern recalled, "You could see rings of vegetation around the wet depressions indicating

Fig. 9-2. Wildin Heritage Prairie mesic area, April 2003, after grazing was stopped the previous fall. Photo by Joe McGovern.

Fig. 9-3. Wildin Heritage Prairie wetland, April 2003, showing effects of previous grazing. Note exposed clumps of hummock sedge and area trampled by livestock. Photo by Joe McGovern.

that native vegetation was present and responding to variations in moisture." The wet margins of the numerous prairie potholes were refuges for some rare and conservative species that favor wet-mesic and wet sites. In addition, some prairie vegetation still remained in the eastern third of the site near the road, which had been more heavily affected by concentrated livestock grazing and trampling by livestock.

In earlier years, the family had harvested prairie hay from the site for decades, thereby retaining most of the prairie species. Jim, one of four Wildin children, recalled the wild hay his father harvested from the site with horses: "You'd see wildflowers in the hay that you wouldn't see anywhere else." Evidence of various native species in the vicinity of the prairie potholes and anecdotal recollections of the Wildin children about unusual wildflowers on the site encouraged the INHF to take a chance and purchase the property. McGovern praised the actions of the family, "Because of their low-impact management approach, the Wildins have preserved part of a disappearing ecosystem: an 80 acre virgin prairie remnant, completely clear of invasive trees and shrubs, located on some of Iowa's best, flat, black soil."

The short-term goal of the INHF was to begin to restore the native prairie community. Nearby remnants such as the Bernau Hay Prairie and the west end of Stinson Prairie were potential restoration reference sites. Grazing was terminated with acquisition of the property to allow the vegetation to recover. The initial response of the vegetation the next spring was disappointing—a growth flush with Kentucky bluegrass, redtop, and dandelions with a scattering of prairie species. However, additional prairie species appeared throughout the summer and early fall.

Thomas Rosburg, INHF advisor and professor of biology at Drake University in Des Moines, inventoried plant species on the site during the 2004 growing season. Rosburg's baseline inventory documented a large number of native plants, 124, including some rare and uncommon species. He noted one species of special concern, hummock sedge, and several species with high coefficients of conservatism, prairie Indian plantain, bottle gentian, white camass, Riddell's goldenrod, and prairie violet.

"Though hurt by intensive grazing, the site probably has more species that will show up when it's had more time to recover," observed Rosburg. "With this kind of diversity, at least the western half of the site is of state preserve quality." McGovern noted, "Although there are prairie species scattered throughout the more disturbed eastern one-third of the site, it is recovering more slowly."

The INHF began to reduce invasive species by having crews of student interns hand-pull sweet clover and Canada thistle. The weeds were hand-pulled,

Fig. 9-4. Recovery of Wildin Heritage Prairie mesic area from grazing, August 2004, the second year after grazing was stopped. Photo by Joe McGovern.

instead of sprayed with herbicides, to avoid the possibility of overspray killing nearby native species. They did spot-spray individual Canada thistle plants growing in fencerows and heavily infested patches near adjacent crop fields. In 2004, the first prescribed burn was conducted on 15 acres of the most degraded part of the remnant near the road. According to McGovern, the fire curtailed growth of some of the nonnative species and stimulated some native species (figs. 9-4 and 9-5).

Long-term goals of the INHF include continuing restoration activities to improve the species composition and vigor of the native prairie community. They may eventually use the site as a source of local seed for other area projects. They plan to increase species richness by collecting seed from native prairie plants in higher-quality areas and interseeding into the more degraded areas with similar plant communities. The seed may be broadcast into the ashes after a prescribed fire.

After reading about Rosburg's baseline inventory of the site, one might form the opinion that Wildin Heritage Prairie is of such high quality that it is not a good example of a degraded prairie remnant in need of restoration. The number of native species is certainly higher than expected from a heavily grazed

Fig. 9-5. Recovery of Wildin Heritage Prairie wetland, 2005, the third year after grazing was stopped. Photo by Joe McGovern.

pasture. However, the east end is greatly degraded, many native species have disappeared from that area, and the abundance of others is greatly reduced. Although Rosburg predicted that the native species count would exceed 124, it is likely that the native species composition prior to settlement exceeded 250. It can be conservatively estimated that haying, grazing, and trampling extirpated at least 50 species. There is much restoration work to be done to put the site on a trajectory toward its presettlement composition. Fortunately, the current condition of the site suggests that it is reasonable to expect such a trajectory (figs. 9-6 and 9-7).

Fig. 9-6. Recovery of Wildin Heritage Prairie mesic area, 2009, in the seventh year after grazing was stopped. Photo by Joe McGovern.

Fig. 9-7. Recovery of Wildin Heritage Prairie mesic area, 2009, in the seventh year after grazing was stopped. Photo by Joe McGovern.

Summary

- Prairie remnants that have been heavily disturbed, modified, and damaged are candidates for this form of restoration.
- Project goals provide the framework for planning, executing the plan, and monitoring the restoration.
- A reference remnant with similar soil, topography, and hydrology will serve as a model for a restoration site.
- Past land use history of the site is very helpful in developing a restoration plan.
- The restoration process includes halting degradation, controlling weedy species, increasing the abundance of existing species, adding species if necessary, containing costs, and protecting rare and endangered species.
- Establish permanent sampling transects and photo points for monitoring and evaluating the restoration project.
- Allow vegetation to recover on heavily grazed sites, and initially remove woody species from brushy sites to open the area to light.
- Initiate prescribed burning as soon as the vegetation fuel load is sufficient to carry fire.
- Costs can be contained with grants from conservation organizations and volunteer workers.
- Insure that the restored site has long-term protection and management.
- Exercise caution in reintroducing extirpated species.
- A restoration is completed when the site is restored to the condition specified by the goals of the plan.

Prairie Management

DARYL SMITH

Overview

Presettlement prairies were shaped and maintained by wildfire, herbivores, and climatic extremes. Today's prairies are mostly small, isolated patches that require active management to maintain their diversity. This chapter focuses on the management techniques needed to maintain the integrity and diversity of prairie remnants, restored remnants, and established reconstructions. Natural resource managers use techniques such as prescribed fire, haying, selective mowing, grazing, brush cutting, and strategic application of herbicides to maintain prairie communities. The utilization of these techniques and their advantages and disadvantages will be discussed.

The Need for Prairie Management

Climate, fire, and grazing were the primary processes that originally maintained the tallgrass prairie. After Euro-Americans entered the tallgrass prairie in the 1830s, they converted most of it to intensive agriculture, replaced bison and elk with cattle and other livestock, suppressed fire, dissected the prairie with roads and railways, and covered it with buildings and concrete. The result is a fragmented ecosystem with a few scattered remnants. The remnants either no longer contain elements of the primary processes of the original ecosystem or lack the capability to respond to those that remain. Due to their small size and relatively extensive edge, they are under constant stress from invasive species, sedimentation, reduced genetic vigor, herbicide drift, nutrient overload, air pollutants, and other human disturbances and are thus more likely to lose critical components of the remaining prairie community. Furthermore, lost species cannot be recruited from other nearby prairie sources, because extensive cultivation of the surrounding areas has eliminated them. Obviously, there is a need for managerial intervention to retain high-quality tallgrass prairie remnants.

We now know that remnants need to be managed to maintain the prairie community and prevent degradation. However, when preservation of prairie began in the 1930s and 1940s, most conservationists believed that management was unnecessary, assuming that disturbed remnants would undergo natural secondary succession and return to their presettlement condition. By the late 1960s, it was becoming apparent that some kind of intervention was needed, because many prairie remnants were becoming degraded from invasion by woody species and nonnative weeds and the loss of conservative prairie species.

Despite a recognized need, many prairie remnants continue to be inadequately managed due to insufficient funds, limited personnel, incomplete understanding of prairie ecosystems, and/or lack of knowledge of the biology of affected prairie species. In the absence of fire and grazing, gray dogwood, smooth sumac, and buckbrush encroach on prairie remnants, form large thickets, and shade out the prairie vegetation. Eastern red cedar seedlings are quite sensitive to fire and are easily killed when small; however, with heavy grazing and an absence of fire, they become major invaders in many areas of the tallgrass prairie. For example, they invade the steep slopes of the Loess Hills prairies bordering the Missouri River and the hill prairies along the Mississippi River, often forming closed-canopy groves.

Management of areas adjacent to remnants is critical to maintaining the components and integrity of the prairie communities. Acquisition and management of buffer lands around prairie remnants will reduce negative impacts from the surrounding altered landscape. Prairie should be reconstructed on these buffer lands with seed from prairie remnants. The buffer lands will also require management, but they protect the biological integrity of the core remnant.

Prairie Management Techniques

A well-designed prairie management plan must include specific goals and management objectives to attain these goals. Maintaining a diverse prairie community capable of responding to environmental changes is a desirable primary goal. Other possible goals could be to replicate a particular type of prairie as determined by climate, soil, and hydrology (for example, blacksoil prairie, sand prairie, or wet prairie) or to maintain a hay meadow as an example of a culturally modified prairie (Howe 1994).

Management for high diversity in prairie remnants and reconstructions is an attempt to replicate historic, large-scale natural disturbances with carefully planned and timed actions. Management techniques include prescribed fire, haying, grazing, selective mowing, woody species removal, and strategic application of herbicides. Knowledge of physiological processes and developmental stages of plants is important to determining appropriate seasonal timing and frequency of techniques used to enhance desired species diversity and control undesirable species. A good prairie management system to increase diversity utilizes techniques that vary by intensity, season, and location. Avoid repetition of a technique in the same location at the same time each year unless you have a different management objective.

PRESCRIBED FIRE

Prescribed fire is an effective tool for prairie management that replaces both lightning and fires ignited by Native Americans. Most prairie managers use fire to minimize litter accumulation, insure continued dominance of warm-season grasses, and maintain species diversity (Lekwa 1984; Bragg 1978; Kucera 1970). It is essential that the use of prescribed fire be based upon well-defined management objectives. Prescribed fire affects vegetation in many different ways. It affects the composition and structure of prairie communities and is essential for maintaining the diversity of native prairie species. It slows the invasion of woody species and stimulates dominant prairie vegetation. Due to their extensive root systems, prairie species are well adapted to fire; likewise, many have perennial buds just beneath the surface of the soil that are protected from fire. Some species with above-ground buds have the ability to resprout after being burned.

Research and experience have demonstrated that fire can be used to effectively manage native and reconstructed prairies. Major benefits include the creation of a favorable microclimate by removing accumulated litter and an increase in light at the soil surface to both increase prairie species and control many nonnative species. Secondary benefits include an increase in flowering and seed production of some native species, a reduction in pathogenic fungi and microbes, the improvement of wildlife habitat, and the regrowth of certain species. If prescribed fire is excluded from small prairie remnants, the prairie vegetation and associated relict species will be lost over time. Immediate post-fire effects on vegetation may be transitory, and definitive changes in the plant community will occur several years later as the result of subtle and indirect effects of the fire on plant physiology and growth.

Strategic burning can be used to address special management problems within limited areas of a prairie. A firebreak is created around a highly degraded area or one infested with invasive species, and the area is then burned with a hot, intense fire to eliminate undesirable species and, ideally, invigorate the prairie species.

Prescribed burn objectives are determined by the management plan for the area being burned. Examples of objectives for prescribed burns are to reduce and remove litter; to curtail the increase of woody species; to stimulate growth of warm-season grasses; to enhance diversity by increasing forbs, sedges, and cool-season grasses; to control exotic species; to control invasive species; to reduce wildfire hazards; to increase native seed production; to improve wildlife habitat; to curtail plant pathogens and control plant diseases; and to prepare sites for reconstruction activities.

Tallgrass prairie is dominated by matrix-forming, warm-season tallgrasses that occupy much of the space in the plant community. Forbs and cool-season grasses occupy the space between larger dominants. Varying the seasonal timing of prescribed fire affects the structure and plant composition of prairie communities. Prairies will burn at almost any time of the year if there is sufficient dead material present (Bragg 1982). Differential responses of plant species to the timing of prescribed fire allow for manipulation of the vegetation to achieve management goals. Late spring fires favor warm-season grasses; dormant-season (early spring and fall) fires benefit perennial forbs, cool-season grasses, and sedges, thereby increasing diversity. Invasive, cool-season grasses such as smooth brome and Kentucky bluegrass are most effectively controlled by late spring prescribed fire (table 10-1).

How species respond to fire is determined by their stage of development and the ways they are affected by the microclimate modification. Plants that are actively growing, flowering, or forming seed are utilizing stored energy reserves. If they are top-killed by prescribed fire at that stage, regrowth uses more stored energy and weakens them. Consequently, frequent fire over time may result in their decline or elimination. On the other hand, dormant plants are unaffected by prescribed fire, and after the burn they take advantage of the lack of competition from other species to grow vigorously (Davison and Kindscher 1999).

Developmental stages of plants vary with latitude; blooming occurs at a later date at higher latitudes. Consequently, calendar dates for prescribed fires vary from south to north. Willson (1992) determined that effective control of smooth brome is achieved when tillers are at the 5-leaf stage. At this stage of development, the impact of fire on the brome plants is most stressful. This stage usually occurs in late spring, but the specific date will vary with latitude and weather.

Prescribed burning of tallgrass prairie is typically done in early spring, although that is usually not the best time from an ecological perspective. The

Table 10-1. Seasonal Effects of Prescribed Fire on Plants

	Dormant-Season Burns			Growing-Season Burns		
Species	Fall	Winter	Early Spring	Mid Spring	Late Spring	Summer
Big bluestem		pos. (6)	pos. (6)	pos. (3, 6)	pos. (6)	neg. (2)
Indian grass		neg. (6)	neg. (6)	pos. (3)	pos. (6)	neg. (2)
Little bluestem	pos. (6)	neg. (6)	pos. (4, 6)	pos. (6)	neg. (4, 6)	
June grass		pos. (6)	pos. (6)			
Sedges		pos. (6)	pos. (6)	pos. (5, 6)	neg. (6)	
Perennial forbs	pos. (3)	pos. (6)	pos. (3, 6)	neg. (6)	neg. (6)	pos. (1, 2)
Leadplant		pos. (6)	pos. (6)	pos. (6)	neg. (6)	
Kentucky bluegrass		neg. (6)	neg. (6)	neg. (6)	neg. (6)	
Smooth brome			pos. (8)		neg. (7)	

Note: pos. = positive, neg. = negative.

Sources: (1) Biondini et al. 1989, (2) Ewing and Engle 1988, (3) Gibson 1989, (4) Henderson 1992, (5) Nagel 1983, (6) Towne and Owensby 1984, (7) Willson and Stubbendieck 1997, (8) personal observation by Daryl Smith.

rationale for burning at this season is usually attributed to one or more of the following: ease of igniting vegetation, cultural traditions, and a desire to remove the previous year's growth and stimulate green-up. Varying the seasons of prescribed fire will elicit differing responses from plants and increase diversity. Late spring is an appropriate time to burn if the management objective is to increase warm-season grasses such as big bluestem and Indian grass. However, repeated annual burning in late spring will likely create a stand dominated by big bluestem and Indian grass with few forbs and native cool-season grasses. Very little research has been done on the effects of late summer fire, even though such fires occurred regularly in presettlement times. We recommend that prairie managers develop fire regimes that vary the seasonal timing of prescribed burns as well as the frequency of the burns.

Fire frequency is the return interval between fire events on a given site; it is one of the major influences on species composition of tallgrass prairie. Annual and biennial spring fires favor warm-season grasses, while longer intervals favor forbs (Kucera 1990; Hulbert 1986). Determining the fire frequency to use for a site depends on how often it is necessary to burn to achieve fire management goals or to replicate the fire-return intervals of the presettlement ecosystem. Presettlement fire-return intervals have to be inferred from fire scars and growth rings of trees in nearby savanna communities or from pollen cores. Based upon available evidence, an average fire-return interval of 3 to 5 years is commonly accepted, but estimates range from 2 to 10 years (Cohen 1998; Wright and Bailey 1982). Fires were likely more frequent during drought periods and less frequent when the climate was moister and cooler.

Prairie managers usually use the accepted fire-return interval of 3 to 5 years for prescribed burning. The interval between fires is usually shortened for wet sites (2 to 3 years) and lengthened for drier sites (6 to 8 years). Warm-season grasses decline with less frequent burning, while forbs, cool-season grasses, annuals, and woody species increase. The result is higher species diversity, but some of the increase may include undesirable invasive species. Optimal fire frequency is site-specific, and monitoring may be necessary to insure the proper frequency for a particular site. For example, a 27-year study (Kucera 1990) of spring burning on a prairie in north-central Missouri showed that the greatest diversity occurred with fire intervals between 2 to 5 years. However, at Konza Prairie in eastern Kansas, species diversity continued to increase up to 7 years after burning (Knapp et al. 1992).

Trained personnel should develop the burn plan for the site and conduct the prescribed burn (fig. 10-1). Each prescribed burn should be carried out only under specified conditions of wind speed, wind direction, temperature, and humidity (the prescribed unit burn plan). Such burns are carefully planned with firebreaks established ahead of time. If smoke-sensitive areas are in the vicinity of the burn site, the prescribed fire plan should address smoke management. Not only is public health a concern, but smoke can also impair visibility in residential areas and along highways. Prescribed burning should be conducted only when the atmosphere is slightly unstable, to allow for maximum lift of the smoke column. Burn bosses should be familiar with techniques that reduce or control the amount of smoke and direct it away from smoke-sensitive areas. Modifying ignition patterns, time of day of the burn, and location and size of the burn units and making allowances for wind direction can reduce the amount of smoke produced.

Although prescribed fire is an essential management tool, it is not a cure-all for tallgrass prairie management problems. To use prescribed fire appropriately, do not burn prairies indiscriminately without management objectives that take into account all species — both plant and animal — of the prairie community. Since most of the total number of animals are insects, the well-being of that group should be considered in the formation of the management plan.

Fig. 10-1. Igniting a flank fire for a prescribed burn. Photo by Daryl Smith.

Many entomologists recommend caution because some prairie-restricted insect species are fire-sensitive (Hessel 1954; Riechert and Reeder 1970; Opler 1981; Schweitzer 1985). Indiscriminate burning without concern for fire-sensitive organisms could eliminate relict populations of some insects that inhabit today's small, fragmented prairie remnants. Insects that overwinter in litter or plant stems are particularly vulnerable to fire. Burning the entire site could endanger some insect populations, particularly those that tend to be distributed in patches across the prairie remnant. Managers can minimize insect losses by restoring habitat diversity, eliminating invasive landscape features such as tree- or shrub-lined trails, and burning with restraint (Panzer 1988).

Data indicate that prescribed burning of portions of prairie remnants at intervals of 2 to 5 years does not negatively affect most prairie-inhabiting insect species. For those affected by fire, recovery is generally rapid (Panzer 1988). Additional data suggest that fire management has contributed to the protection of several species that otherwise would have been lost (Panzer and Schwartz 2000). In spite of expressed concerns, there is no compelling reason to cease using prescribed fire in prairie management, although care should be taken not to be overzealous in its use.

Grassland birds have variable responses to prescribed burning, but over the long term many species benefit from 3-to-5-year burn rotations. Burns scheduled in the spring (March to late April) and in the fall (September to November) outside the summer breeding season are generally best for grassland birds (Sample and Mossman 1997).

We recommend that no more than half of a prairie remnant be burned at any one time. For large remnants, burning a third of the site is more appropriate. With a 3-year burn frequency, burn a third of the site each year. If possible, avoid burning all of a similar habitat at the same time. Don't reignite skipped areas. A mosaic of unburned patches is a typical postfire pattern and provides refugia for animals.

HAYING

Harvesting of prairie or "wild" hay to feed livestock in the winter was a common practice for a century or more after Euro-American settlement. Prairie hay was, and still is, highly valued as horse hay because of its high palatability. Prairie hayfields were usually harvested once in midsummer and were occasionally lightly grazed. Some farmers maintained the prairie-haying tradition well into the twentieth century. Most of the larger blacksoil prairie remnants that survived to modern time were maintained as prairie hayfields. By being managed as hayfields, they were coincidentally managed as prairies, with the haying partially substituting for the burning and grazing that had maintained the presettlement prairie vegetation.

Floristic studies of a variety of grasslands in the tallgrass prairie region of Kansas by Jog and colleagues (2006) showed that warm-season prairie hay meadows possessed the highest species richness and also contained a number of highly conservative native species. The researchers noted that hay meadows are important islands of plant biodiversity and can support threatened and endangered plants. They also indicated that these remnants are one of the few remaining reservoirs of native prairie species and recommended their protection in perpetuity.

Haying is an especially effective prairie management technique when used in conjunction with prescribed fire and/or grazing. Like fire, haying is very effective in reducing litter accumulation and creating a favorable microclimate for prairie species. By cutting and removing the plant canopy, haying increases the light intensity at ground level. This provides an opportunity for recruitment and establishment of new seedlings to increase species richness and diversity. In some ways, the effect of haying mimics that of a summer fire.

Haying can be used to simulate fire effects when climate conditions are not suitable for burning. If fire is not permitted in an urban location, cutting and removing the prairie species as in haying are probably the best alternative. In some instances, it may be desirable to use haying to maintain a prairie hayfield to continue a long-term cultural tradition and to provide an example of the benefit of long-term haying.

When using haying to manage for increased diversity, the following techniques are recommended. Use a hay/no hay rotation, and hay less than half of the site each year—that is, a 3-year rotation with one-third of the site being hayed each year. Harvest hay in midsummer (late July or early August), leaving a 6- to 8-inch stubble, and leave vegetation regrowth for winter and nesting cover. For optimal diversity, use prescribed fire on a 3- to 4-year rotation in conjunction with biennial haying. If cattle are available, consider using a biennial plan that alternates grazing and haying. The timing and frequency of the haying, burning, and grazing should be carefully planned and varied as much as possible. Avoid haying or burning when grassland birds are nesting.

Hayed prairie meadows are very attractive and species-diverse; they contain a high ratio of forbs to grasses. Haying is usually done in midsummer just before the dominant warm-season grasses begin to flower. Haying at this time or later stresses the actively growing dominant grasses, allowing the forbs a competitive advantage and opening the area for seedling establishment. Many important forbs decrease in abundance and vigor under early summer haying (early July) but increase with midsummer (early August) or late summer (mid September) haying (Conard 1954).

The species composition of hayed prairie relics is probably not representative of presettlement prairie. Scattered, cultivated hay species such as timothy, sweet clovers, and red clover are characteristic of hayed prairies. These species were likely carried in on haying equipment. The timing of haying affects the species composition of the prairie as well as the yield and quality of the forage. Plants like warm-season grasses that are hayed while flowering are adversely affected. Regrowth further reduces their carbohydrate reserves, and they are unable to compete effectively with species less affected by the haying. Consequently, repeated annual haying at about the same time increases the probability that these stressed species will be reduced or ultimately eliminated. As a result, after several years, some species in the prairie community could be underrepresented, while others could be missing altogether. Species that flower and set seed prior to the haying time or afterward are favored and may increase in abundance. In addition, late summer haying may further weaken some native species by not allowing them sufficient time, especially for warm-season grasses, to replenish carbohydrate reserves.

A large amount of nutrients are removed by haying. Plants of annually hayed prairies tend to be smaller than those of other prairie remnants. Removal of nutrients may benefit native species more than nonnatives, as they are better adapted to low-nutrient conditions than are weedy species. Hayed prairies have less standing dead plant material and fewer apparent woody species. Annual haying effectively curtails woody species by preventing trees or shrubs from reaching any significant size, but it does not eliminate them. For example, haying suppressed quaking aspen growth on wet-mesic portions of Hayden Prairie in Howard County, Iowa, and Clay Prairie in Butler County, Iowa. However, the aspen persisted and quickly became prominent when haying was stopped.

The nesting of grassland birds can be adversely affected by haying. Cutting after July 15 will allow most species an opportunity to raise at least one brood. Mowing the site from the center out will allow escape routes for fledglings. If late-nesting species like dickcissels and sedge wrens are of concern, haying should be delayed until early August (Sample and Mossman 1997).

Like many prairie management techniques, haying requires special equipment and is labor-intensive. Furthermore, modern haying equipment can compact prairie soil and form ruts in wet areas. To reduce the possibility of this type of damage, haying should be done when the ground is dry and firm. Another potential problem is that seed of nonnative species from previously hayed sites can be retained on the equipment. Before haying, inspect the equipment off site and thoroughly clean it to avoid introducing nonnative species to the remnant being managed.

Prairie management for peak hay yields and/or peak nutritive value for forage production is not consistent with managing for maximum native species richness and diversity. "Managing Missouri's Hay Prairies" by the Missouri Department of Conservation is a good reference for hay production.

SELECTIVE MOWING

Selective mowing at a certain time or in a specific location can be very useful in controlling undesirable woody and weedy species. Spot mowing prior to flowering will reduce or eliminate seed production of invasive species such as Canada thistle. Mowing a weedy patch and allowing some regrowth of the target species prior to herbicide application will increase their exposure to the herbicide. Mowing patches of certain shrubs in the summer after they resprout following a prescribed burn will further deplete their carbohydrate reserves and reduce their ability to compete. In addition, late summer or early fall mowing will decrease the frost resistance of the resprouting stems of undesirable shrub species.

Mowing can also be used to promote desirable species. Howe (1999) found that golden alexanders doubled in abundance and increased flowering percentage when mowed in August rather than May. In addition, selective seasonal mowing can be used to reduce dominant species and allow less competitive species to express themselves. For example, slender prairie bush clover, a federally threatened species, doesn't compete well with more dominant and vigorous native species. Selective mowing in the area of populations of slender prairie bush clover in early summer will set back the more dominant vigorous species and reduce competition for the bush clover.

GRAZING

Moderate grazing by cattle can be a valuable tool for prairie management in areas that can accommodate cattle. All sizes of remnants can benefit from grazing, but it is easier to manage cattle on larger tracts. Cattle can be moved on and off the prairie to manipulate and regulate the timing and effects of grazing (Helzer 2001). Ungrazed tallgrass prairie communities are often less diverse than moderately grazed ones, because dominant species form dense stands that limit species richness and structural variation.

Grazing was a key component of the presettlement prairie ecosystem; it influenced the vegetation composition and structure and created a variety of habitats and niches. The result was higher species diversity, patchy distribution of species, and greater variation in vegetation height. Most, if not all, of the tallgrass prairie developed under grazing by bison and elk, although limited information is available regarding the presence and ecological role of these large herbivores east of the Mississippi. However, both bison and elk disappeared shortly after settlement, and cattle became the primary grassland grazers. Within a short time, cattle were confined in fenced pastures and no longer allowed to range freely as they grazed.

Many prairie conservationists have a negative opinion of cattle grazing on prairie. They feel that it degrades the prairie vegetation and is particularly harmful to conservative forb species. This perspective was formed from observations of the poor quality of small, overgrazed pastures in the eastern portion of the tallgrass prairie (Nyboer 1981). This perception, reinforced by continued overgrazing in many areas, has slowed the acceptance of cattle grazing as a management tool. Helzer and Steuter (2005) note that poor-quality pastures are the result of long-term, continuous overgrazing, not cattle grazing in general.

Prairies can be effectively managed with grazing by native and/or domestic herbivores if careful attention is given to the type of grazers, the number per unit area, and the length of grazing time. Helzer (2001) maintains that cattle grazing can be one of the most flexible tools for prairie management, because it provides so many options for regulating the intensity and timing of the disturbance of the prairie. Grazing should be varied in frequency and intensity for the same reasons as fire. By varying timing and intensity, grazing can be used to selectively reduce certain species, allowing other species to gain a competitive edge. The challenge is to select schemes that reduce dominance of some species without eliminating those species more sensitive to grazing. In addition, cattle can be used to contain the growth of woody plants that compete with grasses and forbs. Cattle can be moved to a remnant for short periods of time and/or moved around the remnant by installing temporary electric fences. If there is a possibility that cattle will import nonnative species, they should be quarantined and fed native or seed-free forage until their digestive tracts are cleared of any undesirable seed. Measures should be taken to prevent cattle from creating permanent trails or major loafing areas around water sources and mineral blocks.

The Nature Conservancy uses short-term cattle grazing as part of its management of Broken Kettle Preserve in the Loess Hills in northwest Iowa (Moats 2006). The goal of the grazing management is to supplement the use of prescribed fire and increase biodiversity. Specific objectives include reducing cool-season grasses, maintaining a diverse vegetative structure, providing follow-up to fire in reducing targeted species, and controlling the growth of woody plants. A prescribed fire removes or top-kills brush, increases light at ground level, and encourages the growth of fuel for future fires. Cattle are attracted to recently burned areas and graze the sprouting grass and woody stems. Subsequent prescribed fire may then eliminate woody species stunted or weakened by moderate to heavy grazing.

Scott Moats (2006), Broken Kettle Preserve manager, uses a conservative approach to grazing management and maintains the number of cattle well below recommended stocking rates. He closely observes changes in vegetation to determine when to remove the cattle. Decisions whether to increase or decrease the number of cattle for the next grazing season are also based upon these observations. There is little or no impact from cattle confinement, because the grazing areas are large and the cattle are moved frequently to limit the amount of time they spend in a particular grazing area. Portable electric fencing is used to create paddocks to facilitate the rotational grazing. It is not necessary to quarantine the cattle to prevent the introduction of weedy species, because they are locally owned and have been grazing similar plant communities.

In summary, grazing enhances species diversity by opening spaces for less competitive species. Lack of grazing reduces diversity due to development of a thick canopy of vegetation and a thick layer of litter. Furthermore, light to moderate levels of grazing in the tallgrass prairie increase species richness, compared to heavy grazing or no grazing at all.

It is generally assumed that bison are more suitable than cattle for prairie management because they co-evolved with the prairie. Due to overstocking and the resulting overgrazing, cattle are often characterized as destructive agents responsible for destabilizing grassland ecosystems, while bison are portrayed as more intrinsically beneficial (fig. 10-2).

However, bison are not readily available, are more difficult to handle, and require special fences. Cattle are more common, more manageable, and easier to transport from place to place. Furthermore, measurable differences between bison-grazed and cattle-grazed pastures of the tallgrass prairie may be less than once believed. Towne and colleagues (2005) compared bison-grazed communities with cattle-grazed communities for 10 years at Konza Prairie. There were definite differences between the vegetation in bison-grazed and cattle-grazed pastures, but overall they were 85 percent similar through time. The researchers suggested that differences in how the herbivores are managed may play a larger role in their impact on prairie vegetation than differences in how they graze. The similarity of the grazed vegetation with both species is certainly sufficient to justify the use of cattle in management of tallgrass prairie.

COMBINED FIRE AND GRAZING

Fire and grazing by large ungulates were essential ecological processes in the maintenance of the structure and function of prairies prior to settlement. Although the two clearly influence one another, they have traditionally been viewed as separate disturbances (Collins 2000; Fuhlendorf and Engle 2001). The combined use of fire and grazing, patch-burn grazing, on large prairie remnants provides a method for simulating presettlement disturbance regimes in prairie. The method was first employed with bison on prairie areas larger than 5,000 acres. More recently, it has been shown to be effective with cattle on smaller sites (Davison and Kindscher 1999; Fuhlendorf and Engle 2004; Helzer and Steuter 2005). With this fire-grazing interaction method, a different part of the prairie is burned each year. As indicated in the section on prescribed fire, the fire-frequency schedule is derived from an estimated presettlement fire-return interval. After the burn, grazers are allowed access to the entire area. The grazers are attracted to the burned areas and graze them intensely for a period of time. The grazing temporarily decreases tallgrasses and increases forbs. Diversity increases in these grazed patches. The grazers spend less time on patches burned previously and little or no time on unburned patches. In the

Male bobolink on compass plant,
Doolittle Prairie, Story County, Iowa.
Photo by Carl Kurtz.

Blackmun Prairie, Butler County, Iowa, after a spring burn. Photo by Greg Houseal.

Blackmun Prairie, Butler County, Iowa, in fall after a spring burn. This historically grazed prairie, mostly grazed out in the bottomland, is recovering on slopes where grazing by cattle was less intense. Photo by Greg Houseal.

White sage and prairie cord grass,
Cayler Prairie State Preserve, Dickinson
County, Iowa. Photo by Carl Kurtz.

(*Above*) Young bur oak and white wild indigo in reconstructed tallgrass prairie, Prairie Creek Wildlife Refuge, Marshall County, Iowa. Photo by Carl Kurtz.

(*Below*) Male common yellowthroat on compass plant amid showy tick trefoil, Prairie Creek Wildlife Refuge, Marshall County, Iowa. Photo by Carl Kurtz.

Katydid on rosinweed. Photo by Kirk Henderson.

Iowa's state flower, the prairie rose. Photo by Kirk Henderson.

Compass plants, gray-headed coneflowers, common milkweed, and big bluestem, Prairie Creek Wildlife Refuge, Marshall County, Iowa. Photo by Carl Kurtz.

(*Above*) Purple prairie clover, rosinweed, and showy tick trefoil in reconstructed prairie in late July, Prairie Creek Wildlife Refuge, Marshall County, Iowa. Photo by Carl Kurtz.

(*Below*) Gray-headed coneflowers in Osheim's third-year prairie reconstruction near Story City, Story County, Iowa. Photo by Carl Kurtz.

(*Above*) Marsh phlox and sedges, wet prairie swale, Worth County, Iowa. Photo by Greg Houseal.

(*Right*) Infrared aerial photograph, 2002, with Hayden Prairie State Preserve, Howard County, Iowa, in spring at the center. Vigorously growing vegetation — cool-season pasture, hayfields, and waterways — appears red; tilled agricultural fields appear gray. Pinkish tan indicates dormant vegetation, potentially warm-season native grasses. The blackened southwest portion of the photo and triangular area within the preserve indicate a very recent prescribed burn. Photo courtesy of the Iowa State University Geographic Information Systems Support and Research Facility.

(*Above*) The second growing season for a Mahaska County, Iowa, roadside planting. Canada wild rye is prominent at this stage. Photo by Kirk Henderson.

(*Left*) Butterfly milkweed. Photo by Daryl Smith.

Pale purple cone-flowers, spiked lobelia, and marbleseed, Blackmun Prairie, Butler County, Iowa. Photo by Greg Houseal.

Wild bergamot stands out in a 10-mile planting in Winnebago County, Iowa. Photo by Kirk Henderson.

Fragrant coneflower.
Photo by Dave Williams.

Hoary vervain, black-eyed Susan, and daisy fleabane in a reconstructed pasture with a number of prairie remnants, Spring Hill, Marshall County, Iowa. Photo by Carl Kurtz.

Walkingstick on leadplant, Doolittle Prairie, Story County, Iowa.
Photo by Carl Kurtz.

Fig. 10-2. Using bison for prairie management marks the return of one of the major grazers to its original habitat. Photo courtesy of New Light Media.

absence of grazing, tallgrasses gain dominance and litter accumulates, limiting light and reducing the abundance and diversity of forbs. The result is a shifting mosaic of patches of differing diversities, maximizing plant species diversity over the entire prairie.

Bison and cattle tend to prefer grasses to forbs, thus reducing the matrix of dominant grass species and allowing other species to increase. Therefore, after a burn, the intense grazing opens space between the dominant grasses for new growth of forbs, especially annuals and biennials. The short-lived forbs dominate the patch during that year and the next and then slowly decline under competition from recovering perennials. As a result of the preference of grazers for recently burned areas, fuel accumulates in the unburned and less recently burned areas. The greater fuel load in the unburned area increases the probability and intensity of fire. When an area burns, the patch-burn sequence begins again.

Maintaining high plant diversity in restored and remnant prairies is a constant challenge. A patch-dynamic approach to management of the tallgrass prairie provides a means to develop and maintain a highly diverse, constantly shifting mosaic landscape. Fuhlendorf and Engle (2001) indicate that optimal patch size and fire-return intervals depend upon rangeland management objectives and the amount of time required for burned patches to recover. They suggest that a possible patch-burn management plan for tallgrass prairie could involve burning one-third of a pasture each year—half of the third in the spring and the other half in the summer. This would provide an interval between burns of 3 years.

CONTROL OF WOODY SPECIES

Woody species control is an ongoing management issue for prairie remnants and long-term prairie reconstructions. Historically, eastern red cedar, dogwood, sumac, and other woody plant species were natural parts of the prairie edge. Their populations would grow or shrink depending on the frequency and intensity of fires and/or the extent of browsing by deer, elk, and other animals (Helzer 2001).

Early studies suggested that prescribed fire was a good management tool for preventing woody plants from invading tallgrass prairie (Bragg and Hulbert 1976; Kucera 1960). Frequent prescribed fire at 3-year intervals or less is effective at eliminating seedlings and young saplings in prairies. For example, a well-timed prescribed head fire pushed by wind of 10 to 15 miles per hour in sufficient fuel can remove all red cedar trees and other species under 6 to 8 feet in height. Late summer fires in very hot, dry years may be intense enough to effectively reduce the number and cover of all woody species.

Elimination of woody species that are well established on tallgrass prairies is very difficult. Recent studies of woody invasion at Konza Prairie (Heisler et al. 2003; Briggs and Gibson 1992; Briggs et al. 2002; Briggs et al. 2005) have raised questions about the adequacy of prescribed fire and grazing in controlling established trees and shrubs. Periods without fire provided time for recruitment and establishment of new individuals of shrub species. Annual spring fires prevented recruitment of new shrubs, but shrub cover still increased slightly. Once established, the shrubs persisted, even in frequently burned areas, and shrub cover increased regardless of fire frequency (Heisler et al. 2003). Long-term (20-year) data sets derived from watershed study sites on Konza Prairie were used to track the responses of larger trees and shrubs to different fire frequencies and grazing (Briggs and Gibson 1992; Briggs et al. 2002). Briggs and colleagues (2002) found that woody plant density increased two- to ten-fold in all watersheds except those burned annually. Intermediate (4-year interval) fire frequency actually increased rough-leaved dogwood more than infrequent fire (15-year interval). After bison were added to some of the infrequently and annually burned study sites, woody abundance on those sites increased over that of the unburned study sites. Evidently, the accelerated increase in woody abundance was due to the bison. By reducing the amount of fuel around the trees and removing the low-hanging branches, they prevented fire from affecting the trees.

Invasion of prairies by woody species can be a greater problem than invasion by weeds. Canopies formed by trees and shrubs shade out prairie species and reduce or eliminate the fuel necessary for prescribed fire. Management options include late spring or summer burning, mechanical removal by machines or hand, and treating with herbicides. Hand methods include use of axes, lopping shears, grubbing tools, and chainsaws as well as girdling (fig. 10-3). Fire, mowing, and/or grazing will eliminate many species when they are very young. Late spring burning for 2 to 3 consecutive years is usually the most practical and effective way to control most woody plants. Nonsprouting species such as eastern red cedar can be killed by burning or cutting below green growth. However, sprouting species such as smooth sumac and rough-leaved dogwood require repeated top removal, either mechanical or chemical or a combination of the two. Stage of growth is critical, as cutting at the wrong time stimulates new growth and increases the number of sprouts. Treatment is most effective from full-leaf to early flowering stages when carbohydrate reserves are low.

Most woody species are susceptible to foliar-applied herbicides just after the full-leaf stage in the spring. However, not all species reach the full-leaf stage at

Fig. 10-3. A weed whacker with a circular saw blade is an effective tool for cutting brush. Photo by David O'Shields.

Fig. 10-4. Girdling a tree by removing a strip of bark and cambium can be an effective means of killing the tree by stopping the return of nutrients to the roots and preventing the growth of sucker shoots at the base. Photo by David O'Shields.

the same time. For example, smooth sumac is best controlled with early June mowing, whereas buckbrush should be mowed 3 to 4 weeks earlier. Consecutive top removal for 2 or 3 years at the appropriate time may be necessary to effectively control aggressive species. Larger deciduous trees like Siberian elm, boxelder, honey locust, and green ash require a more individual approach. Girdling — removal of a strip of the cambium layer around the stem — to stop carbohydrate transport from roots to upper parts of the plant is effective in killing trees while avoiding resprouting (fig. 10-4).

Unlike girdling, cutting down the trees is seldom effective alone and needs to be combined with herbicide treatment of the cut stem (stump). Either a concentrated (40 percent) solution of glyphosate or the more toxic Tordon RTU (Ready-to-use) herbicide is effective for stump treatment. Triclopyr can also be used to treat stumps to prevent resprouting. In addition, one chemical form, Garlon 4, can be used to control woody vegetation without cutting by treating a zone of bark along the lower part of the trunk. This basal bark treatment is effective even in winter. Clopyralid (Transline) is unusually effective against plants of the legume family and can be used to control black locust. As discussed in chapter 6, herbicide effectiveness depends upon using the proper chemical at the correct time and rate. Label information on herbicides can be obtained at the following website: http://www.cdms.net.

Selecting a woody control method depends on the species involved, degree of invasion, topography, economics, adjacent land use, and management objectives. If several different woody species are involved, combinations of methods like cutting and grazing or burning and cutting are less expensive and more effective than a single treatment method (Towne and Ohlenbusch 1992). The key to woody species management is to recognize potential problems and control them before they become severe. Once they are under control, more general management techniques may suffice.

CONTROL OF INDIVIDUAL SPECIES

It may be necessary to target individual species that are particularly abundant or invasive because general management techniques may be ineffective or inadvertently promote some species. The Nature Conservancy has an excellent website with comprehensive information about the control of weed species; see http://tncweeds.ucdavis.edu/control.htm.

As an example of individual species control, white sweet clover, a biennial, will increase with spring burning at 3-year intervals. A spring burn will stimulate a flush of germination within the seed bank. These plants mature, produce seed the second year, and replenish the seed bank before the third-year burn. Kline (1986) determined that a combination of an April burn followed by a May burn the next year was most successful in reducing white sweet clover. The April burn stimulated a flush of germination and first-year growth of white sweet clover. The May burn the next year killed the biennial plants before they flowered. Kline found that mowing in July as a follow-up treatment the second year instead of burning in May was not as effective, but it did reduce sweet clover substantially. On thick stands, repeating the back-to-back (April and May) burn treatment separated by 2 years of not burning almost completely removed the white sweet clover.

Controlling invasive species is an ongoing process. Effective control of a species requires an understanding of its development, physiology, and habitat requirements. If invasive species are removed, desirable species should be available to replace them. A number of organizations and agencies have pertinent information on the control of invasive species. See the Center for Invasive Plant Management, the USGS Northern Prairie Wildlife Research Center, the USDA Management Plans by Species and the National Invasive Species Information Center, and the Illinois Natural History Survey.

Summary

- Fire, grazing, and climate were the primary factors that maintained the original tallgrass prairie.
- Human settlement fragmented the tallgrass prairie and eliminated wildfire and large herbivores, resulting in cessation of these ecosystem processes.
- Lack of prairie management results in degradation and loss of prairie remnants.
- Prairie management techniques include prescribed fire, haying, grazing, selective mowing, woody species removal, and selective application of herbicides.
- Prescribed fire is an effective tool for prairie management that replaces both lightning and fires ignited by Native Americans.
- Major benefits of prescribed fire include control of many nonnative species, removal of accumulated litter, increased light at soil surface, and an increase in species diversity.
- Varying seasonal timing of prescribed fire is the best way to maintain native plant diversity.
- A fire frequency of 3 to 5 years is commonly used to minimize litter accumulation, insure continued dominance of warm-season grasses, and maintain species diversity.
- Although prescribed fire is an essential management tool, it is not a cure-all for tallgrass prairie management.
- Care should be taken to avoid the elimination of native insect species that inhabit remnants.
- Haying is an especially effective prairie management technique when used in conjunction with prescribed fire and/or grazing.
- Hayed prairie meadows are attractive and species-diverse and contain a high ratio of forbs to grasses.
- When using haying to manage for increased diversity, hay less than 50 percent of the site with a hay/no hay rotation; hay in late July or early August, leaving a 6- to 8-inch stubble; and leave regrowth for winter and nesting cover.
- Frequent mowing in the early stage of prairie reconstruction increases the establishment of prairie species by preventing the formation of a weed canopy and increasing light at ground level.
- Selective mowing can be used to control undesirable woody and weedy species and can be used to reduce dominant species to allow the expression of less competitive species.

- Grazing as a management tool can enhance species diversity by opening spaces for less competitive species and preventing the development of a thick canopy and litter layer.
- Effective prairie management by grazing with native and/or domestic herbivores requires careful management of the type of grazers, the number per unit area, and their length of time on the remnant.
- An interactive method of fire and grazing known as patch-burn grazing can be effective in creating high plant diversity in a mosaic of patches across the prairie landscape.
- Effective control of invasive species necessitates an understanding of their development, physiology, and habitat requirements.
- Prescribed fire and grazing can effectively prevent the encroachment of woody species into tallgrass prairie.
- Control of woody species that have established on a prairie remnant or reconstruction requires more aggressive and intensive management techniques, including the use of herbicides.
- Individual species may have to be targeted for specific management techniques if they become particularly abundant or invasive.

Special Cases

Prairie in Public Places

DAVE WILLIAMS

Overview

Prairie reconstruction in public spaces challenges our perception of urban landscaping. These plantings are quite different from traditional "neat and tidy" landscapes. Prairie reconstruction projects can be very visible and controversial in any community and can elicit strong public responses, both pro and con. This chapter deals with how to positively influence public perceptions. The success of public plantings is tied to three key principles: communication, planning, and implementation.

Two case studies have been included in this chapter to illustrate the process of planting prairie in public spaces. The first case study involves planting a prairie in a city park. This project was initiated by a very influential person in the community and also required the partnering of city and state personnel, a local service club, and students from a state university and community college. The second case study involves planting a prairie on the campus of a state university. This project was proposed as a means to reduce maintenance costs. The project required the cooperation and support of administrators and grounds personnel and illustrates the value of taking ownership in the project.

Communication

Reconstructing a prairie in an urban setting is less about the project development and implementation and more about politics. Unfavorable public perception can sink the best project. Most of us are comfortable with mowed turf grass and the tidy look of vegetation in public spaces. When confronted with something different, like a reconstructed prairie of tallgrasses and other unfamiliar plants, many people tend to react negatively. It is understandable that these prairie plantings are sometimes perceived as weed patches and associated with an unmanaged landscape (Hough 2004).

Communication with elected officials and public employees should occur *before* initiating any prairie reconstruction project. This is important for two reasons. First, informing and educating public officials about a prairie project will give them ammunition when they respond to public criticism. Second, most cities have ordinances that regulate planting within the city limits. Since these ordinances are general and open to interpretation, they can potentially be exploited by individuals against alternative forms of landscaping. When public officials have been educated about the value of planting prairie, some of the concerns expressed by the public can be mitigated before they become explosive topics at a city council meeting. Better yet, before the first prairie project is drawn up, consider working with the parks department to request an addition to the city code that would allow for prairie reconstructions and plantings of native grasses and wildflowers within the city limits.

The code of ordinances for the city of Cedar Falls, Iowa, for example, prohibits the following "noxious weeds": "Quack grass (*Agropyron repens*), Perennial sow thistle (*Sonchus arvensis*), European morning glory and field bindwood (*Convolvulus arvensis*), Horse nettle (*Solanum carolinense*), Leafy spurge (*Euphorbia esula*), Perennial peppergrass (*Lepidium draba*), Russian knapweed (*Centaurea repens*), Buckthorn (*Rhamnus*, not to include *Rhamnus frangula*), and all other species of thistles belonging in genera of *Cirsium* and *Carduus*, Butterprint (*Abuilon theophrasti*), Cocklebur (*Xanthium commune*), Wild mustard (*Brassica arvensis*), Wild carrot (*Daucus carota*), Buckhorn (*Plantago lanceolata*), Sheep sorrel (*Rumex acetosella*), Sour dock (*Rumex crispus*), Smooth dock (*Rumex altissimus*), Poison hemlock (*Conium maculatum*), Wild sunflower (wild strain of *Helianthus annus* L.), Puncture vine (*Trimbulus terrestris*), annual Teasel (*Dipasacus* spp.), Grass exceeding 12 inches in height, Wild vines or wild bushes." However, exceptions include the following areas: "(1) Prairie grass areas, wildflower planting areas, natural reserve and preserve areas, urban woodlots, wildlife refuge and conservation areas, wetlands and natural waterways, all as recognized and identified by a governmental agency. (2) Land zoned agricultural under the zoning ordinance of the city exceeding five acres in size. (3) Other conservation or natural areas deemed appropriate by the city council after consultation with the director of human and leisure services or his/her designee."

Similarly, the "natural lawn ordinance" of Madison, Wisconsin, states that "lawns shall be maintained to a height not to exceed eight (8) inches in length." However, "Any owner or operator of land in the City of Madison may apply for approval of a land management plan for a natural lawn, one where the grasses exceed eight (8) inches in height, with the Department of Planning and Development." This land management plan is "a written plan relating to management of the lawn which contains a legal description of the lawn upon which the grass will exceed eight (8) inches in length, a statement of intent and purpose for the lawn, a general description of the vegetation types, plants, and plant succession involved, and the specific management and maintenance techniques to be employed. The management plan must include provisions for cutting at a length not greater than eight (8) inches the terrace area, that portion between the sidewalk and the street or a strip not less than four (4) feet adjacent to the street where there is no sidewalk, and at least a three (3) foot strip adjacent to neighboring property lines unless waived by the abutting property owner on the side so affected."

It is important to communicate with residents, businesses, and organizations affected by any native planting before it appears on the landscape. The support of many people can be gained simply by telling them what a prairie is and what to expect (Maag 1994). The message needs to be positive and to extol the benefits of the project. Depending upon the type or scope of the project, there can be many benefits to promote: improving water infiltration and water quality, reducing dependence upon fossil fuels (by decreasing use of fertilizers, herbicides, and mowing), eliminating use of irrigation, introducing a butterfly habitat, providing an educational opportunity for the public, creating a green image, or restoring a historic landscape (Mariner et al. 1997).

Managing public expectations through open communication will result in positive perceptions concerning the project. The public has been trained to expect instant gratification from public landscaping projects. Through press releases, public presentations, and one-on-one dialogues, project coordinators can convey that it takes a few years for a native planting to develop. The weedy appearance is only temporary until the native perennials take over.

Reconstructing an urban prairie can generate a lot of public interest, and newspapers will usually run stories if contacted. To insure accuracy, provide a summary sheet of the reconstruction project. Include positive statements, quotes from local supporters, and suggestions about how the public will benefit from the new landscape so that the reporter can insert these points into the story.

Internal communication is important, too. There are often several layers of people between the person who developed the prairie plan and the person who has to maintain the planting. Sharing the prairie plan with everyone involved in the project can create advocacy and keep everyone informed at all levels of the project. Project coordinators can use the prairie reconstruction as a tool to enhance communication and job satisfaction among field staff. Site preparation, seeding, and postseeding maintenance (prescribed burning, plant identification, weed control, and woody plant control) are all skills associated with reconstructing a prairie. According to Paul Meyermann, landscape architect for the University of Northern Iowa, "Planting prairie is not easier—it's a different way to do your job that will require working at a higher level." Providing key staff with the opportunity to learn some of these skills, such as prescribed burning, can increase receptivity to planting and managing prairie.

In every community, local advocates for natural landscaping and respected influential people—including members of service organizations, school personnel, and elected officials—can play a critical role in the project's success. Use their influence to garner support for the project. Selling the project to these folks will make it much easier to sell it to the general public. They may even generate financial support for the project.

Planning

A good prairie plan is the foundation of a successful prairie planting. Seek advice from prairie experts when planning a prairie reconstruction. Prairie experts can be found in state and federal agencies, nonprofit conservation organizations, university or community college botany departments, and local parks departments. Experts can anticipate project needs unique to the site, assist with species selection, help avoid pitfalls when working with contractors, and inspect the project to insure that tasks have been done correctly.

Not every public area is appropriate for a native planting. Consider the intended use for the site. High-use areas usually require frequent mowing. Unmowed taller vegetation is appropriate for lower-use edge areas. Golfers, for instance, are more likely to accept a reconstructed prairie on the edge of a golf course, which does not interfere with their game, rather than one adjacent to a fairway.

With the guidance of a prairie expert, develop a written plan. This takes the guesswork out of the project for the contractor or maintenance crew. List any specialized equipment, seed, and herbicides that will be used. Provide a clear description of each task and a timetable of when each task is to be completed. Even a short plan will reduce errors or miscommunications and provide something tangible for the stakeholders to support and promote.

The plan should include maintenance tasks specific to reconstructed prairies such as developing a seed mix, prescribed burning, control of noxious weeds, and removal of unwanted woody plants. These maintenance tasks require specialized training and certification. Determine up front who is responsible for maintenance and who will pay for it. Your state's Department of Natural Resources, land grant state college extension offices, and online contractor sites should have cost information associated with such activities.

The Department of Transportation for your state will ask private contractors to comply with particular specifications. The Iowa Department of Transportation, for example, requires private native-seeding companies that contract with the department to follow these specifications for roadside spraying, native grass seeding, and mowing. For roadside spraying, "The Contractor shall furnish and apply (Glyphosate) a broad spectrum non-selective systemic herbicide in April/May on the existing vegetation that's actively growing in the native grass seeding areas and the area receiving mulch for existing and proposed plants. This work shall be done after vegetation regrowth of the first mowing. The rate of glyphosate and surfactant shall be according to manufacturer's label recommendations. Herbicide should be applied when wind conditions are 10 mph or less. Contractor shall not spray area 7' adjacent to the rock shoulder, nor 2' from the right of way line/fence. The herbicide shall be applied by a certified pesticide applicator, Category 6 right of way.

"Existing trees and shrubs in these areas shall not be sprayed or damaged during this contract work. Where existing trees and shrubs are present, the contractor shall make application as close as spraying and seeding equipment allows without damaging the trees and shrubs. Subsequent applications: Subsequent applications of herbicides shall be applied in 10 day intervals to vegetation not killed. After complete kill of the vegetation, the area shall be seeded." For native grass seeding, the "contractor shall apply seed between May 01–June 30. Seeding shall begin 7' adjacent to the shoulder (delineator posts), and end 2' from the right of way fence, unless noted otherwise.

"The Contracting Authority will furnish the seed. The University of Northern Iowa Tallgrass Prairie Center will mix the seed and store it at the center. The contractor shall coordinate timing of pickup of seed with the Tallgrass Prairie Center for the project. The Contractor shall not add or include any other seed for the project. The Contractor shall apply the native seed with a Truax drill or equivalent. The seed must be accurately metered and uniformly mixed in each box during drilling. The Contractor shall place the seed in the drill box labeled on the seed bag. The drill shall contain an aggressive picker wheel for continual mixing of seed. The drill shall be equipped with disc furrow openers and packer assembly to compact the soil directly over the drill rows. The drill shall be equipped with a no-till attachment. The drive wheel shall maintain ground contact at all times. The Contractor shall not apply seed in wet soil conditions that would cause the seed to be placed deeper than specified. Planting depth shall be ⅛ inch. Reseeding of these areas will be required at the Contractor's expense if damages to these areas occur due to the Contractor's negligence during the establishment period of the contract. Fungicide is not required."

Mowing "includes 7 mowings of native grass seeding areas. One mowing shall be accomplished prior to spraying (initial mowing— 6 inch height). Three mowings at four week intervals, beginning 30 days after seeding (6 inch height). Three mowings at four week intervals the following year, beginning the first two weeks in May, June, and August. Mow in May at the 6 inch height and all other mowings at 11 inch height."

Provide sufficient space between buildings and the planting. The dense, fine fuel of prairie vegetation, when burned, creates intense radiant heat. Infrared waves from radiant heat can peel paint, melt vinyl siding, and, in extreme cases, spontaneously ignite wood siding.

Smoke produced by the fire should also be considered. Smoke over roadways is extremely dangerous. Visibility can be reduced to almost zero in heavy smoke. Smoke can also enter through a building's intake ducts and cause interior damage. When prescribed fire cannot be used, other methods should be implemented to manage the native vegetation (see chapter 10).

No matter where the planting is located, it should be accessible to the public. According to Meyermann, "A planted prairie has to be a place that people are encouraged to visit; otherwise it doesn't have value to them." Trails are community assets; they improve the quality of life. A mowed trail meandering through the planting invites people to use it. Providing signs along the trail gives the public another visual cue that it's okay to go into the planting. People are always curious about the individual plants they discover in natural landscapes. Staking a small sign near an individual plant that gives its common name and some interesting characteristics will satisfy their curiosity.

Implementation

A number of things can be done during a prairie reconstruction project that will improve the quality of the final product. The project leader should have some familiarity with planting natives. This person can communicate with the contractors to insure that specifications of the reconstruction plan are followed. If bad weather doesn't allow the project tasks to be completed according to the plan, an experienced project leader can make appropriate adjustments in the timetable to best fit with a native-seeding project.

Inspect the work as it is being done, not after it is completed. Contaminated seeding equipment, improper seeding depth, uneven seeding, and improper seed mixing are all factors that will negatively affect the quality of the project. These problems can be detected and prevented by inspecting the operation while the work is being done.

Require documentation from the contractor. If herbicides are used on the project, ask to examine the spray records. This will provide verification that the proper chemical was used, mixed at the right concentration, applied by a licensed person, and applied at the correct time (fig. 11-1). The seeding contractor should submit seed-test labels for every species used for the project; this verifies that the seed quality and quantity specified in the plan were used for the project.

Case Study 1: Rotary Prairie Planting at Big Woods Park

In the fall of 2001, Marvin Diemer, a retired politician and an Iowa Rotarian, approached the Tallgrass Prairie Center at the University of Northern Iowa (UNI-TPC) about planting a prairie at Big Woods Park in Cedar Falls. Diemer had been inspired by a prairie reconstruction, Wilson Prairie in the nearby town of Traer, and the educational opportunities it offered the residents of the community. The Tallgrass Prairie Center provided the technical support, oversight, and expertise necessary to make the prairie restoration a success.

Part of Big Woods Park had been used as a borrow area for a nearby highway construction project and then deeded to the city of Cedar Falls. The park is within the city limits and is surrounded by residential and commercial development. Staff from the UNI-TPC met with the manager of the Parks and Recreation Department and the city planner to determine where a prairie could be planted and to get their blessing for the project. The city of Cedar Falls had developed a long-range recreation and maintenance plan for Big Woods Park. A 10-acre area within the park had been designated for development of a campground. City officials agreed that the development of a campground was probably not going to happen and decided that planting a prairie on the site seemed to be a good alternative. A paved bicycle trail bisecting the 10-acre planting site was a bonus feature. The trail is part of an extensive network of interconnected bicycle trails in metropolitan Cedar Falls and Waterloo (fig. 11-2).

A landscaping plan for a tallgrass prairie reconstruction of 10 acres was drafted by the UNI-TPC staff. What made this reconstruction project unique was the diverse seed mix used for the project. The soil at Big Woods Park is excessively well-drained sandy loam. Cedar Hills Sand Prairie, one of the highest-quality sand prairie remnants in the state, is located within 10 miles of Cedar Falls. It was used as a model to develop a seed mix for the Big Woods planting. The seed mix developed for this project included 85 species of grasses, forbs, rushes, and sedges. Since many of the older Rotarians were retired farmers, the high cost of seed for this project was a tough sell for Diemer. He convinced the

Fig. 11-1. Sample pesticide application record.

Certified Applicator ______________________________

Certification ____________________ Phone ____________

Customer ______________________________

Address ____________________ Phone ____________

Crop or other area treated:

Number of acres of other units treated:

Pest(s) controlled and developmental stage:

Severity of infestation, infection, etc.:

Stage of crop growth (if applicable):

Date and time pesticide was applied:

Soil condition (if applicable):

(wet, dry, cloddy)

Temperature _____ Humidity _____ Cloud cover _____

Wind direction and speed _____

Pesticide used (name of product and formulation):

How pesticide mixed (if applicable):

(active ingredient per gallon, etc.)

Total amount of pesticide applied:

(gallons/lb acre or other unit treated)

Specific location of application (if applicable):

Township:

Section:

Quarter:

Map of treated area (if applicable):

Fig. 11-2. The city of Cedar Falls maintains a mowed trail through the 10-acre Big Woods Park prairie restoration. Photo by David O'Shields.

Rotarians that this seed mix was needed for this planting to be as good as the one in Traer. The Rotarians funded the entire project; most of the funds went toward the cost of the seed.

The first order of business was to remove thousands of Siberian elm trees growing on the site. In the plan, all nonnative trees were to be hand-cut, with the stumps treated with an herbicide and removed from the site. Removing the trees with a bulldozer was not an option, due to the potential for soil erosion into a lake. This was the first test of the project. The UNI-TPC staff contacted the city, but their maintenance crews were extremely reluctant to hand-cut the trees, partly due to a perceived public backlash against extensive tree cutting in a park. Rather than putting the city into a difficult position, a staff person from the UNI-TPC showed the Rotarians how to safely cut the trees and treat the stumps with an herbicide. A press release in the *Waterloo–Cedar Falls Courier* described the prairie that was scheduled to be planted at Big Woods Park and indicated that removal of some nonnative trees was needed on the planting site:

> In May of this year, 10 acres of native Iowa grasses and wildflowers will be planted along the bike trail at Big Woods Lake in Cedar Falls. Cedar Falls

> Rotary purchased the prairie seed and has been working on the site removing non-native trees and brush to prepare the site for seeding.
>
> "This is the highest quality prairie planting that I have ever been involved with," said Dave Williams of the UNI Tallgrass Prairie Center. "More than 80 species of wildflowers, grasses and sedges will be planted at a cost of $15,000.00, which will be paid entirely by the Cedar Falls Rotary," Williams said.
>
> After seeding, the planting site will be mowed for the entire summer to reduce weeds and to increase prairie plant establishment. Big Woods Rotary Prairie will provide outdoor enthusiasts an opportunity to see a spectacular assortment of prairie grasses and wildflowers and the experience of being on the prairie.
>
> Future projects on the site include a savannah planting and construction of an interpretive trail.

In 3 days, the Rotarians had gathered enough volunteers to complete the work. They quickly cut and piled the trees and treated the stumps (fig. 11-3). City crews then chipped the trees and hauled the woodchips away. In retrospect, the project benefited from the city crews' reluctance to hand-cut the trees. The tree cutting gave the Rotarians ownership of the project. It became their planting, and they were committed to doing whatever it took to make the project succeed. When other difficult tasks arose over the years, such as hand-pulling sweet clovers, Diemer gathered the volunteers and they got the job done.

Seeding the site was an event. Most of the members of the Rotary Club showed up to seed. Carrying gallon ice cream buckets filled with a mixture of small seed and sand, they seeded the entire 10 acres by hand, while the UNI-TPC staff seeded the larger seed with a no-till drill. Rotarians were very interested in learning how the no-till seed drill planted the seed. The *Waterloo–Cedar Falls Courier* sent a reporter and a photographer to cover the story. The front page of the Sunday edition featured a story of the prairie seeding at Big Woods Park accompanied by a photograph of two prominent Rotarians hand-seeding: Jon Crews, the mayor of Cedar Falls, and Joel Haack, dean of the College of Natural Sciences at UNI.

Diemer has been the driving force behind this prairie project. He has energized the Rotarians to build an information kiosk, convinced the city to mow a permanent trail through the prairie, persuaded the local utility company to fund the planting of 12 large oak trees, organized crews of Rotarians to assist in weedy species control, and influenced other citizens to write letters of support for the project. In the fall of 2006, Diemer and a large group of Rotarians, for

Fig. 11-3. As part of the site preparation before seeding prairie at Big Woods Park, the Cedar Falls Rotarians cut and piled nonnative trees. Photo courtesy of Marvin Diemer.

the first time since the project began, collected seed off the planting. The seed will be used for a 3-acre addition to the prairie. The land for the addition was offered by the city without being asked.

Case Study 2: Natural Landscaping at the University of Northern Iowa

In 1993, UNI adopted a master landscape maintenance plan. This plan classified levels of maintenance of all grass areas on campus. In 2000 and 2001, fuel prices were high, and UNI experienced significant budget cuts. The plan allowed UNI landscape architect Paul Meyermann to experiment with natural landscaping features to reduce mowing costs. He selected grass areas that were low use but frequently mowed. Meyermann believed that converting low-use areas into prairie would encourage people to visit them, thereby adding value to these areas.

In 2001, a half-acre area adjacent to the UNI Museum was selected for a natural landscape project. The museum staff was adamantly opposed to a prairie planting, because they felt that this would take away space that they needed for outdoor functions. Meyermann presented a drawing of his vision to the museum

Fig. 11-4. Prairie planting at the University of Northern Iowa Museum in Cedar Falls. The center of the planting has been left as mowed turf grass so the museum can utilize this space for events. Photo by David O'Shields.

staff and assured them that the maintenance staff could easily mow the prairie if the museum needed the space for a weekend function. This was an acceptable compromise. The prairie at the museum was planted, and subsequent prairie plantings were made around the original planting over the next 5 years. Today, there is a half-acre semicircle of mowed turf grass in the middle, surrounded by diverse tallgrass prairie (fig. 11-4). Most of the prairie plants are perennials that can easily rebound from one mowing during the growing season. Trails meander throughout the prairie and are connected to the mowed turf grass core. Museum staff use the area as an outdoor classroom. It was never necessary to mow the prairie. Meyermann stated, "If we can take some of the knowledge, skills, strengths, and patterns that we recognize in the prairie and use them to our best advantage for people, then we are contributing something."

Meyermann has taken a proactive approach to educating his ground crews. In the spring of 2006, the UNI-TPC staff held a workshop on prescribed burning for Meyermann and his maintenance crews. For most, this was their first introduction to TPC staff. They learned about the nuts and bolts of establishing native vegetation, site preparation, seeding, initial postseeding maintenance, prescribed burning, and weed control. The TPC staff demonstrated the equipment used for prescribed burning. The workshop ended with the maintenance crews conducting prescribed burns of three campus plantings, supervised by the TPC staff.

Public land often has transitional functions that can radically change over time. According to Meyermann, "Most landscape installations in our environment are good for 5 to 10 years, and after that they may have to survive having utilities trenched through them or being used as a staging area" (2006). In 2003, UNI constructed new athletic fields and had 3 acres of the perimeter planted to prairie. In the summer of 2004, a very large Morton building was constructed in the middle of this prairie planting. The area around the building was also disturbed for construction of an access road. Before construction, there were no noxious weeds in the planting. However, after construction, Canada thistle invaded the site, requiring multiple chemical applications to eradicate. Some prairie wildflowers were inadvertently killed as a result of spraying. However, the remainder of this planting is progressing well.

Summary

- Keep public officials informed about the project.
- Inform the public *before* planting, and tell them what to expect.
- Convey to everyone that a prairie reconstruction takes a few years to develop.
- To insure accuracy, submit a fact sheet to the media.
- Develop a written prairie reconstruction plan.
- Be specific in the plan about special equipment, herbicides, tasks, and timetables.
- Distribute the prairie plan to all personnel involved with the project.
- Provide key staff with specialized training associated with reconstructing a prairie.
- Promote the benefits of using native vegetation.
- Find a local champion to support the project.
- Seek advice from prairie experts when planning a reconstructed prairie.
- Budget for specific tasks such as prescribed burning, mowed firebreaks, and control of noxious weeds and unwanted woody plants.
- Provide sufficient space between buildings and the planting.
- Make the prairie accessible to the public.
- Designate a project leader familiar with planting natives.
- Inspect the work as it is being done, not after it is completed.
- Require documentation from the contractor about spray records and seed-test labels.

12 Roadsides and Other Erodible Sites

KIRK HENDERSON

Overview

Conditions found in roadside rights-of-way create significant challenges for prairie reconstruction. Slopes are often too steep for tractor and drill. Compacted soil requires more aggressive site preparation. Steep slopes and concentrated water flow increase the need for soil stabilization. And proximity to motor vehicle traffic makes it harder to manage the plantings once they are established. Roadside prairie reconstruction efforts are further complicated by federal storm water regulations, public opinion, and local politics. To our credit, thousands of acres of roadside rights-of-way have already been restored to native vegetation. The benefits of native plants provide plenty of incentive for overcoming the challenges. These benefits range from the functional and environmental to the aesthetic and cultural. Of all the benefits, none is more crucial than the conservation of soil and water. When it comes to the use of native prairie species as a means to protect and enhance our natural resources, we are just getting started.

The Advantages of Native Prairie Vegetation for Roadsides

State and county road departments choose to work with native vegetation for many reasons. As durable, long-lived perennials, prairie plants are very competitive and occupy the root zone to such an extent that weeds are deprived of necessary water, nutrients, and space to grow. Native prairie species perform well in poor soil. Their extensive root systems penetrate 6 to 8 feet deep, conditioning and rebuilding soil that has been disturbed and compacted during road construction. Their deep, fibrous roots are especially effective at holding soil and stabilizing slopes. Deep roots also enable prairie plants to survive drought and high concentrations of salt.

A wide swath of prairie grass can trap blowing snow and thereby increase the storage capacity of the ditch and reduce the amount of snow deposited on the road surface. Native plantings have even been known to reduce whiteout during blizzards. In most cases, a late-season shoulder mowing is all that's necessary to prevent the tallgrasses from causing additional snow to be deposited on the road surface.

Sturdy native plants stand up to storm water runoff, slowing it down and allowing more rain to soak into the ground. Storm water infiltration is further increased, because native plants make the soil surface spongier by increasing organic matter in the upper layer. The leaf surface of an acre of prairie grass can hold 50 tons of water droplets, reducing runoff by allowing water to evaporate or drip gradually onto the soil (Weaver 1954). And the natural decay of deep roots leaves channels through which water can percolate deep into the soil.

Colorful wildflowers and the rich fall colors of prairie grasses add natural beauty to the right-of-way. Diverse native plantings provide valuable food and cover for songbirds and game birds, which is especially valuable in intensively farmed regions where habitat for these species is scarce. And for many people, restoring a piece of our natural heritage is incentive enough for planting tallgrass prairie species along our roadsides.

Whether the goal is protecting groundwater through ecologically sound management practices, saving money with more self-sustaining, better-adapted species, or simply making better use of the vast acreage devoted to roadsides, there is great value in native prairie vegetation.

Planting Site Considerations

Talk to landowners before planting. If private landowners plan to mow the right-of-way in front of their home or business, plant turf grass. Make a pitch for natives, but be prepared to give them what they want. When native vegetation is used within city limits, city officials must be brought on board ahead of time. You will need their support to defend the planting if constituents demand a more manicured appearance. Make sure the mayor, city council, and parks board support the project and understand that native plantings take time. Accompany them to native plantings in the area and instruct them in prairie appreciation. In some situations, tall vegetation growing too close to the road might limit a motorist's view. In most cases, shoulder mowing alone provides plenty of visibility. Beyond the shoulder, the foreslope usually takes the taller grasses down and out of the way of sightlines. Roadside managers maintain visibility at intersections and driveways by planting a mix of short grasses such as brome, fescue, and perennial rye along the edge of road shoulders.

If reed canary grass, crown vetch, bird's-foot trefoil, or other such invasive species occupy a site being considered for a native planting, choose a different site. These species are hard to eliminate, even with repeated herbicide applications. They can appear to be gone only to reclaim the site within a few years, starting from a few surviving plants or from seed present in the soil.

Some highway rights-of-way already contain native plant populations. These precious remnants of the original prairie landscape somehow survived road construction or moved back into the right-of-way after the road was built. A prairie remnant can be valued for possessing just a few species of note or as a diverse plant community. Big or small, prairie remnants provide a glimpse of the past and are valued as sources of genetic material and models for prairie restoration. They all merit protection.

Look for prairie remnants where an old railroad right-of-way parallels the highway or where land may have been too rocky or too wet to till (fig. 12-1). A thorough inventory of roadsides in your jurisdiction is the best way to document the location of remnants and prevent their destruction in the future. Generally, do not try to enhance a remnant by interseeding it with native seed unless that seed comes from remnants in the immediate vicinity.

Fig. 12-1. This high-quality prairie remnant was spared the plow because of the limestone outcropping and because it is sandwiched between highway and railroad rights-of-way. Photo by Kirk Henderson.

Designing a Seeding Mix

The goal is to create a strong plant community by combining sufficient diversity of native species, employing both grasses and forbs. Select species that play different roles and fill different ecological niches within the planting. And combine them in such proportions that they remain in balance. These are lofty goals. Fortunately, most native prairie plants are long-lived perennials that provide a great deal of stability and continue to pay dividends year after year, regardless of the inadequacies of our design. Functional groups of prairie plant species—quick-establishing species, warm-season grasses, cool-season species, legumes, crowd pleasers, spring beauties, wet ones, little guys, prairie regulars, and gilded lilies—are identified below. Include species from each category to achieve a well-rounded mix with all the benefits that native vegetation has to offer.

Some native species develop faster than others. They are especially important in roadside plantings, because they provide early erosion control and good public relations color. Examples of these quick-establishing species are Canada wild rye, black-eyed Susan, tall dropseed, and partridge pea.

Roadside plantings rely heavily on warm-season grasses, prominent members of the plant community. Like corn (another warm-season grass), these grasses continue to grow throughout the summer. They provide long-term erosion control as well as good color in fall. Examples are big bluestem, Indian grass, switchgrass, and side-oats grama (fig. 12-2).

Plantings are further strengthened by the addition of cool-season species that start growing earlier in the spring. These plants provide important erosion control in late winter and early spring plantings and, ideally, occupy the niche sought by nonnative grasses such as smooth brome. Examples are Canada wild rye and Virginia wild rye, western wheatgrass, and most sedges.

The prairie flora includes many legumes that thrive in roadside plantings. These mostly nitrogen-fixing plants establish readily and provide food and cover valuable to wildlife. Examples are white wild indigo, round-headed bush clover, showy tick trefoil, and milk vetch (fig. 12-3).

People like dramatic color. The subtle beauty of native plantings can go unnoticed by the traveling public. Species in this crowd-pleasing category create masses of color noticeable at 65 miles per hour. They establish readily and have some of the least expensive seed. Examples are gray-headed coneflower, ox-eye sunflower, wild bergamot, and New England aster (fig. 12-4).

It's relatively easy to extend the progression of blooming color well into the fall. Spring color, on the other hand, is harder to come by. The following plants

Fig. 12-2. (*Above*) Warm-season grasses play an important role in roadside seeding mixes, stabilizing long, steep slopes and providing long-term erosion control. Photo by Kirk Henderson.

Fig. 12-3. (*Right*) Native legumes improve habitat by providing structure, canopy layering, and food vital to wildlife. Photo by Kirk Henderson.

Fig. 12-4. Wildflowers such as gray-headed coneflower and wild bergamot can be quite showy, creating masses of color noticeable even at 65 miles per hour. Photo by Kirk Henderson.

start blooming in mid spring, about the earliest that we can expect much prairie color to be visible from the road. Examples are Ohio spiderwort, golden alexanders, large-flowered beardtongue, and cream false indigo.

There is a tendency to load the mix with upland species. A complete roadside mix includes prairie species for moist bottoms, too. Examples of these wet ones are bluejoint grass, swamp milkweed, dark green bulrush, and sneezeweed.

Like the kids chosen last when picking teams on the playground, plants with inconspicuous flowers are often overlooked. Including a few of these little guys is good for team chemistry. Examples are common mountain mint, bastard toadflax, alumroot, wild quinine, white sage, and of course sedges. Most of us think life is too short to worry about a bunch of sedges. However, you'll do well to get to know a few of these long-neglected plants such as Bicknell's sedge, shortbeak sedge, oval-leaf sedge, and brown fox sedge.

Don't overlook popular species that did not show up in the other categories. They are the prairie regulars, plants we always look for that make us feel at home on the prairie. Everybody knows their names: compass plant, butterfly milkweed, rattlesnake master, and prairie blazing star (fig. 12-5).

Fig. 12-5. This native planting has matured beyond the showier, early phase and now features some familiar, long-lived community members such as prairie blazing star and rattlesnake master. Photo by Kirk Henderson.

Beware of investing heavily in forb species that cost too much and/or may not succeed under roadside conditions. At the same time, even limited success with these gilded lilies can prove very satisfying. Experiment with small amounts. Examples are shooting star, prairie violet, Michigan lily, prairie phlox, and bottle gentian.

A short section of road can present a wide range of growing conditions, from gravelly, well-drained soil at the top of the slope to heavy, often-saturated soil at ditch bottom. For this reason, when working in especially narrow rights-of-way, it is more efficient to plant one mix that includes species for a wide range of habitats. Apply the same mix over the entire area, and let it sort itself out. Wider rights-of-way may have wet areas or dry areas large enough to justify designing and planting seed mixes specific to those spots.

Resist the temptation to plant nonnative grasses or legumes with the native species. They may help stabilize the site early on, but they will provide so much competition and persist to such a degree that in the end the native component will suffer, and the integrity of the planting will be compromised. Examples are tall fescue, perennial rye, crownvetch, red clover, and bird's-foot trefoil. In fact, don't plant bird's-foot trefoil or crown vetch anywhere near a native planting.

Omitting the taller native grasses as a safety precaution limits the adaptability of the mix. Shorter native grasses such as little bluestem, side-oats grama, and tall dropseed do best in sandy, well-drained soil. Little bluestem might eventually fill in and cover the slope from top to bottom. This will take several years. Even when the intent is to allow forbs to show through, include at least half a pound to 1 pound of big bluestem per acre, and increase the little bluestem rate to 4 or 5 pounds per acre. See table 12-1 for a sample seed mix of tallgrass species native to much of the Upper Midwest.

The reintroduction of native vegetation is crucial to the conservation of our soil and water resources. Never feel guilty about using public funds to pursue this goal. Do feel guilty about failures that result from cutting corners. Purchase good-quality seed of known origin. Invest in proper storage facilities so seed viability remains high. Don't hire seeding contractors who bid so low that they cannot do the job right.

Seeding Rates

Steeper slopes require heavier seeding rates. To stretch limited budgets and provide necessary erosion control, roadside seeding mixes rely heavily on native grasses and generally contain a higher proportion of grass seed compared to forb seed.

We recommend the following minimum seeding rates for grasses and forbs combined: 40 seeds per square foot on level sites, 60 seeds per square foot on 3:1 slopes, and 80 seeds per square foot on 2:1 slopes.

In roadsides, where seeding rates for native grasses may run as high as 15 to 20 pounds per acre, it is difficult to keep grasses from overwhelming forbs. A mix that is 25 percent forb seed is considered minimal for achieving any degree of diversity. We recommend the following minimum ratios of forbs to grasses: 10 forb seeds and 30 grass seeds per square foot on level sites (approximately 10 pounds per acre), 15 forb and 45 grass seeds per square foot on 3:1 slopes (approximately 14 pounds per acre), and 20 forb and 60 grass seeds per square foot on 2:1 slopes (approximately 18 pounds per acre).

Cover Crops

Cover crops are fast-growing, usually nonnative species used to hold the soil until the native species get established. The objective is to select a cover crop that will reduce soil erosion without providing too much competition for native seedlings. Preferred cover crop species have an upright growth habit, are

Table 12-1. Sample Seeding Mix

Species	Seeds/ Ounce	Pounds/ Acre	Seeds/ Square Foot	Functional Category
Grasses				
Big bluestem	10,000	1	3.7	warm-season
Little bluestem	15,000	2	11	warm-season
Indian grass	12,000	1	4.4	warm-season
Side-oats grama	6,000	2	4.4	warm-season/ quick-establishing
Tall dropseed	30,000	0.5	5.5	warm-season/ quick-establishing
Canada wild rye	5,200	1	2	cool-season/ quick-establishing
TOTAL		7.5	31	

Species	Seeds/ Ounce	Ounces/ Acre	Seeds/ Square Foot	Functional Category
Sedges				
Brown fox sedge	100,000	1	2.3	wet one/little guy/ cool-season
Shortbeak sedge	29,000	1	0.7	little guy/ cool-season
TOTAL		2	3	
Forbs				
Partridge pea	2,700	8	0.5	quick-establishing/ legume
White wild indigo	1,700	1	0.04	legume
Purple prairie clover	18,000	1	0.41	legume
Showy tick trefoil	5,500	2	0.25	legume
Milk vetch	17,500	1	0.4	legume
Round-headed bush clover	8,500	2	0.38	legume
Leadplant	16,000	1	0.37	legume
Golden alexanders	11,000	2	0.5	spring beauty
Ohio spiderwort	8,000	2	0.37	spring beauty
Large-flowered beardtongue	14,000	1	0.32	spring beauty
Gray-headed coneflower	35,000	1	0.8	crowd pleaser
Black-eyed Susan	92,000	0.5	1	crowd pleaser/ quick-establishing
New England aster	67,000	0.25	0.38	crowd pleaser
Ox-eye sunflower	6,300	2	0.29	crowd pleaser
Wild bergamot	75,000	0.25	0.43	crowd pleaser
Stiff goldenrod	70,000	0.25	0.4	prairie regular
Butterfly milkweed	3,825	2	0.18	prairie regular
Pale purple coneflower	5,300	2	0.25	prairie regular

Table 12-1. (*Continued*)

Species	Seeds/ Ounce	Ounces/ Acre	Seeds/ Square Foot	Functional Category
Rattlesnake master	7,500	2	0.35	prairie regular
Compass plant	660	3	0.05	prairie regular
Foxglove beardtongue	130,000	0.125	0.37	prairie regular
Prairie blazing star	11,000	2	0.5	prairie regular/ wet one
Sneezeweed	130,000	0.25	0.75	wet one
Blue vervain	93,000	0.25	0.53	wet one
Culver's root	800,000	0.063	1.15	wet one
Swamp milkweed	4,800	4	0.44	wet one
Common mountain mint	220,000	0.125	0.63	wet one/little guy
TOTAL		41 oz	12	
Total seeds per square foot = 46				

Note: Species shown are native to much of the upper midwestern tallgrass prairie. This mix would not be appropriate for use in all parts of the tallgrass prairie region.

short-lived, and do not form a dense canopy. Besides holding the soil, cover crops help reduce drying by sun and wind. Cover crops are recommended for slopes that are 3:1 or greater.

Oats and winter wheat are excellent cover crops (fig. 12-6, table 12-2). They grow rapidly in cool weather, they withstand moderate frost, and their seed is relatively inexpensive. Numerous varieties of winter wheat can overwinter in Iowa, especially with good snow cover; oats do not overwinter.

For native plantings, winter wheat is preferred over winter rye. Winter rye is taller, more persistent, and possibly allelopathic—chemically inhibiting the growth of native forbs. Perennial nonnative grasses and legumes are not recommended for use as cover crops. They provide too much competition early on and remain too much a part of the planting.

When cover crops are planted with the native species, they are called nurse crops or companion crops. Recommended seeding rates for nurse crops are as follows: in spring, 1.5 bushels of oats or 1 bushel of oats and 5 pounds of annual rye per acre; in summer, 2 bushels of oats or 1 bushel of oats and 10 pounds of annual rye per acre; in fall, 20 pounds of winter wheat per acre. When cover crops are planted alone, pending a more favorable time to establish natives, they are called temporary seedings or stabilizer crops. Recommended seeding rates for temporary seedings are as follows. In summer, plant 30 pounds of oats,

Fig. 12-6. This stand of oats provides temporary cover, holding the soil pending establishment of the slower-growing native species. Photo by Kirk Henderson.

10 pounds of annual rye, and one of the following warm-season species: 5 pounds of piper sudan per acre, 10 pounds of millet (Japanese or Pearl variety, first frost halts development) per acre, or 30 pounds of sorghum (grain or forage). In fall, plant 20 pounds of annual rye or 25 pounds of winter wheat per acre.

Mowing the nurse crop before it forms a dense canopy and before it produces a seedhead will promote the growth of the native species. Winter wheat nurse crops must be mowed two or three times the following spring prior to seedhead emergence to prevent seed production and reduce long-term persistence.

When used as a temporary seeding, winter wheat must be destroyed the following spring with herbicides, or it must be mowed two or three times to prevent seed production. Temporary seedings planted during the summer should include a warm-season annual component such as piper sudan grass, millet, or sorghum. Do not seed these species too heavily. One good rain can cause mass germination. Piper sudan may cause concern among landowners as it is sometimes confused with shatter cane.

Table 12-2. Cover Crop Conversion Chart

Species	Pounds/Bushel	Seeds/Ounce	@ 1 Bushel/Acre
Oats	32	910	10 seeds/square foot
Winter wheat	60	937	20 seeds/square foot
Annual rye (*Lolium multiflorum*)	—	12,710	10 pounds/acre = 46 seeds/square foot

Seeding Times in Order of Preference

Generally speaking, mid spring provides the best soil temperature and soil moisture conditions for germination and survival of warm-season species. As stated earlier, roadside seeding mixes tend to be dominated by warm-season grasses.

Early spring is a good time to plant if site conditions permit, even though it might be a few weeks before warm-season grasses start to grow. In early spring, be sure to include a nurse crop of cool-season species such as oats. As summer approaches and rainfall becomes less consistent, conditions for seeding become less favorable.

In the Upper Midwest, the time for dormant seeding begins in November and continues until ice and snow cover the ground. Erodible sites should be mulched if they were not previously stabilized with a temporary seeding. Winter provides a natural cold treatment that improves the germination rate of many forb species (Packard and Mutel 2005). Increase seeding rates for native grasses by 25 percent to allow for losses due to deterioration from the elements.

The USDA Natural Resources Conservation Service recognizes a seeding method called frost seeding, which takes place in February and March on seedbeds prepared in the fall. On erodible sites, include a nurse crop of cool-season species. Do not frost-seed on areas covered with ice or snow.

Many road projects are completed and in need of seeding from July through October. To address the need for seeding erodible sites during these less-than-optimum times, two courses of action are offered. The preferred method is to stabilize the site right away with a temporary seeding. Refer to the above recommendations, and select appropriate species. Return to the site in late fall or the following spring to plant the permanent mix of native prairie species. Use the back-up method, going ahead and seeding the natives in midsummer or early fall, when limited time and resources or administrative pressures preclude following the two-step seeding process. This approach is also for use in the summer, when soil moisture and weather trends are encouraging. When planting in midsummer, increase the seeding rate by 25 percent. Include an appropriate nurse crop to reduce erosion and minimize further drying and crusting of the soil. Plant with a native grass drill, so the seed is placed into the soil, where it will stay put and have better access to moisture. Mulch the site with straw, and crimp it into place to preserve soil moisture. And use your best seed. Don't expect seed left over from last year to do well under stressful conditions.

The Planting Process

For a taste of real-world experience, the information in this section is based on an interview with Joe Kooiker, roadside biologist, Story County, Iowa, on March 18, 2005.

Walk the site before planting; look for gullies, culverts, and other hazards. Measure the project site for an accurate determination of the amount of seed needed to cover it. When you are confident that you have brought along the right amount of seed, you are quicker to realize when you're seeding too heavy or too light.

"To prepare the seedbed, level sites can be worked with a chain-tooth harrow. Our tractors do not have dual rear tires. So steep slopes are ripped with a wide-track dozer. If bulldozers leave a rough surface, don't smooth it up until just before planting. The rougher the soil surface, the less it will erode," says Rob Roman, roadside manager, Linn County, Iowa (fig. 12-7).

When working in heavily compacted soil, disk the site to break it up, working it to a depth of 3 inches if possible. It might be necessary to use a heavy disk just to make some grooves in the surface. Ditch clean-outs can leave the soil surface smooth and hard. Nothing will grow under those conditions. At the very least, try to rough it up. A small cultipacker on a 3-point tractor works pretty well in most ditches. If nothing else, the corner of the cultipacker will dig and rip the ground, the tires will spin, and the bottom of the tractor will scuff, creating a rougher surface where the seed can catch and hang on (fig. 12-8).

"For hydroseeding, we prefer the site to be rough and a little soft. We seed immediately after the loader has left, with no additional seedbed preparation. The rough texture keeps the seed in place, and the softness allows for better root penetration. For drill seeding, firmness is the most important factor. It is easy for seed to get buried too deep in soft seedbeds, either during or after planting," says Doug Sheeley, former roadside biologist, Dallas County, Iowa.

For mixing your own seed, a room with a hard, smooth floor works well. Prop the door open wide, turn on the exhaust fan, and wear dust masks and safety glasses. Measure out the seed with a scale, dump it on the floor in big

Fig. 12-7. "Rip it with the dozer." Heavy equipment is used to loosen compacted soil prior to hydroseeding. The bulldozer's low center of gravity makes it ideal for use on steep slopes. Photo by Kirk Henderson.

Fig. 12-8. All-terrain vehicles navigate steep, narrow situations and are especially effective when you are preparing a seedbed in lighter soil. Photo by Kirk Henderson.

piles, and mix it with scoop shovels. Putting the mixed seed in trashcans works well for hauling it to the site. Seed that won't be planted right away must be kept cool and dry.

A good time for hydroseeding is right after it rains. The seed and mulch really stick better on moist soil, and some moisture is captured under the mulch. Do not start hydroseeding if rain is in the immediate forecast. Mulch needs time to set up before it rains. Wet and gooey mulch is more likely to wash away. When planting with a drill, seed before the rain. Drilling is less effective when the depth bands are caked with mud.

Include oats, annual rye, or winter wheat to green up the site right away. This is for the public as much as for erosion control. Highway engineers, farmers, and people in general like to see something green. Sometimes weeds can be a really good nurse crop if kept from getting too tall.

The native grass drill is preferred for seeding on level rights-of-way (fig. 12-9). Drilling is a one-step process, quicker and cheaper than hydroseeding. However, maneuvering a tractor and drill around the numerous silt fences used on some construction projects can be challenging. For uniform coverage, drill seed at a light rate, and go over the area at least twice. Keep going until the seed

Fig. 12-9. Native grass drills do a good job of incorporating seed into the soil. They are especially effective along roadsides that are wider and more level. Photo by Al Ehley.

is gone. Making multiple passes packs the seed in well and creates more little rills that hold seed and interrupt water flow.

Today's clean seed goes through some of the older drills too fast. Use a filler to slow it down. Some options include bulk-harvested seed that has a high percentage of plant material or little bluestem, which is fluffy and typically takes 50 pounds of material to get 25 pounds of pure live seed.

Calibrate the drill in the shop, and set the rate a little lighter than you actually want it. When you plant, the drill bounces around, causing the seed to come through at a heavier rate. As an example, a particular drill set at 6.5 pounds to the acre might actually seed 8 pounds to the acre. Becoming familiar with the equipment makes this easier to gauge.

Forb seed can run through the drill too quickly. For better distribution of this pricier seed, sprinkle the larger and fluffier forb seed across the top of the other fluffy seed in the drill's middle hopper. Add more forb seed every other round or two. Most native seed is small and lacks the energy to emerge if planted deeper than a quarter inch. Broadcast-seeding the finer-seeded species is a way to prevent them from getting buried. Seeding by hand, followed by light raking, works well on smaller projects.

Some people disconnect the lower end of the seed tubes on their native grass drill. This scatters the seed, allowing some to land on the soil surface instead of all of it being placed in the furrow. Opinions vary as to how or whether this should be done. Some people prefer to unhook only every other tube. Others unhook only the tubes coming from the small seed box. All these variations have the same objective: to keep the seed from having to compete for the same narrow space so some seed is not planted too deeply. When you drill with a trash plow attachment, it should just scratch the surface. If it's making furrows, it's planting too deep.

When hydroseeding, the seeding rate depends largely on how fast the driver is going. Familiarity with the equipment is key. Sometimes hydroseeding leaves a shadow, the area behind larger dirt clods that gets no seed. To reduce the shadow and provide more even coverage, try to seed in two passes, one from each direction. Seed lightly so the seeding rate is not doubled. This is accomplished by having the driver go fast, 7 to 8 miles per hour, and adjusting the seeder to reduce the flow. Seed the area farthest from the road first.

Hydroseeding goes faster using a nozzle with a wide spray pattern. This technique applies the mulch in a gentle rain, leaving an even carpet of seed and mulch on the soil surface. On steep slopes, try to embed the seed by using a more concentrated stream and holding the gun at a sharper angle. Since seed

Fig. 12-10. Applying a slurry of mulch and seed with a hydroseeder. Hydroseeding equipment allows roadside managers to seed roadsides without having to drive on them. This is highly advantageous on steep slopes or under muddy conditions. Photo by Kirk Henderson.

may be damaged going through the hydroseeder and some seed may get hung up in the mulch, increase seeding rates by 25 percent.

As far as the advantages of hydroseeding are concerned, steep slopes can be seeded without getting on them with a tractor. In most cases, the entire roadside can be seeded from the shoulder (fig. 12-10). Ditches that are too wet for drilling can be hydroseeded. Hydromulch reduces soil erosion. Colored mulch tells the public that something has been seeded. Hydroseeding eliminates the risk of planting the seed too deeply.

Disadvantages of hydroseeding include the following. It takes extra time to fill the hydroseeder with mulch and water. Mulch is expensive and can double the cost of seeding. Seeding rates must be increased to allow for seed damaged in the mechanics and for seed mortality from getting hung up in the mulch. Native grasses seem to establish more slowly when hydroseeded. Hydroseeding is strictly a bare-ground application. It's harder to control the seeding rate when hydroseeding.

A 1,500-gallon hydroseeder can cover one-third of an acre per load. With a machine of this size, 7 50-pound bales, or 350 pounds of mulch per load, will yield about 1,000 pounds per acre. That is a token amount—just enough mulch to color the ground, show what part has been seeded, and help carry the seed. For greater erosion control on longer, steeper slopes, use up to 2,000 pounds per acre.

There are many kinds of mulch. A high-paper/low-wood blend works well. Definitely do not use all paper or all wood. If the mulch contains too much wood or poor-quality wood, it won't stick. To learn what good wood content feels like, break open the bale, grab a corner, and grind it in your hands. Instead of virgin wood fiber, sometimes ground-up pallets are used.

Mulch that comes in loosely packed 40-pound bales is actually harder to handle than the solid 50-pound bales. The bales are dustier, too. Reduce back injuries by using a loader to get the mulch up on top of the hydroseeder. Also, some mulch comes with packets of dye to be added separately. These dyes are a mess when they fall on the floor and get stepped on. When they hit the agitator, people get dyed green. Buy the predyed mulch.

"Some people recommend hydroseeding the seed first by itself and coming back over that with the hydromulch—the advantage being that all the seed is in contact with the soil, not hung up in the mulch. If you're going to seed in two passes, I think you're better off putting the seed on with a broadcast seeder and hydromulching over that. Seed does not stay suspended in the water tank as well or carry as far without mulch," according to Jeff Chase, roadside manager, Des Moines County, Iowa.

Vicon fertilizer spreaders work great for broadcast-seeding native seed (fig. 12-11). They can go on the tractor, on the trailer, and down the road with a 1-ton pickup. A broadcast seeder on a 3-point tractor is more compact than a drill and better for getting in and out of ditches. You can back right up to silt fences, spread the seed over the top of them, and cover both sides. For very clean seed, the Vicon can be adjusted down to the nth degree, and it has such a good agitation system that the fluffy seed doesn't get stuck. Just open the gate a lot wider for the fluffier seed.

A few species in the seeding mix may have seed so fine that only a handful will be needed for several acres. To improve distribution of such fine seed over a large area, mix it with some kind of carrier. Sand is best. Kitty litter, perlite, or oats can be used also. Mix the seed and carrier in a bucket, and scatter it over the site by hand. Many wet prairie species fall into this category.

Revegetation is the process by which roadsides already planted to traditional cool-season grasses are converted to native prairie vegetation. The site is prepared by killing the existing vegetation with an application of a 2 percent solu-

Fig. 12-11. Broadcast seeding is more effective now that the seed industry is better at cleaning native seed. Seed flows through the hopper easier and is propelled through the air better once its various appendages have been removed. Photo by Kirk Henderson.

tion of glyphosate in April or May. For a better kill, apply the herbicide one or two more times during the season before seeding in November. Apply the herbicide when the existing vegetation is green and growing but no more than 12 inches high.

If herbicides are applied over a long enough period, such as from fall to fall, the old roots break down pretty well. When it's time to plant, so little plant residue remains that the area can be hydroseeded. Otherwise, it's best to plant with a drill. Often, killing the brome releases Canada thistle or other weeds. Keep an eye on the planting, and spot-spray weeds as they appear.

On revegetation sites, do not treat the first 4 feet of the foreslope or mow zone with glyphosate; leave the short, cool-season grasses in place. Keep the entire planting mowed during the first growing season to reduce competition for the new seedlings and to keep the weeds down.

Packing seed tightly to the soil insures a more consistent flow of moisture from the soil to the seed. The result is better germination and better survival of seedlings. You can't hurt the seed by packing it too tightly. Even hydroseeded

Fig. 12-12. A lot of expense goes into building roadside slopes and ditches. Erosion control blankets are a worthwhile investment in erosion-prone situations where steep slopes and concentrated water flow exist. Photo by Kirk Henderson.

plantings benefit from being rolled in (after the hydromulch has dried). This makes them grow much faster — especially the nurse crop. A 4-foot cultipacker section on a 3-point tractor will go places the tractor and drill can't. Rolling in the seed is not so important in the fall, unless the mix includes a nurse crop. Oats and wheat need to be pushed into the soil.

Sites that are highly erodible can benefit from the use of erosion control fabrics. These additional erosion control measures should be considered when back slopes are 2:1 and 15 to 20 feet long. Roadsides that channel runoff from a large watershed are definite candidates for this kind of insurance, as are slopes comprised of loose or sandy soil (fig. 12-12).

Double-netted wood fiber in 4-by-60-foot rolls is an effective product that's easy to handle. On somewhat flatter sites, straw matting, also double-netted, in 15-by-30-foot rolls can be used. The matting helps the seed grow as much as it holds the soil. Seed the site prior to installing the blanket. Remove debris, rocks, clods, and so on to insure that the blanket remains in contact with the soil surface.

On long slopes, run the blankets down the slope. Secure a blanket along the top of the slope by stapling it into the bottom of a trench 6 inches deep. The trench should run along the length of the installation. Fill the trench with compacted soil. Leave the lower end of the blanket unstapled to allow you to overlap the next blanket to form a shingle pattern. Overlap adjacent blankets so the upstream blanket is on top of the next blanket. Use staples at least 6 inches long (sandy or loose soil may require longer staples), and staple the blanket seams and body per manufacturer recommendations.

To install erosion blankets in a channel, excavate a trench along the top of the channel's side slopes and at its upstream end. Roll the blanket downstream and secure its edges in the trench. Center a blanket in the channel's bottom to avoid having a seam under the area of maximum water flow. Leave the end of the blanket unstapled to allow for overlap. Place the downstream blanket under the upstream blanket to form a shingle pattern.

The erosion control industry has produced a wide range of products designed to hold soil in place. Long, steep slopes and concentrated water flow make them a necessity in many roadside plantings. Erosion mats and blankets, hydromulch, straw mulch, and compost add cost to the project; however, the added investment might save the cost of regrading and replanting. And it's hard to put a price on peace of mind.

Maintenance

Native vegetation establishes faster and more successfully when the planting is kept mowed during the first growing season. Most native plants remain quite small throughout the first growing season—most of their energy goes into root growth—and are easily shaded out by dense weeds. Plan on mowing at least three times the first season. Weed growth varies on individual sites and must be closely monitored. By the time weeds are dense and 2 feet tall, some seedlings may already have been lost. Mowing at this late stage drops too much plant material on top of seedlings. Keep a close watch on weed growth. In a wet growing season, a planting might need mowing every 2 weeks.

It's not so important what kind of mower you use. Just mow. Mowing to a height of 4 inches is not so short that it'll hurt plants. But keep in mind, trying to mow that close with a 15-foot bat wing mower will result in some scalping. Scalping definitely hurts plants. On uneven ground, just a tickle of the switch will put the blade in the dirt. Eight inches with no scalping is better. Part of deciding when to mow is guessing when farmers are going to spray their fence rows; mowing the weeds first helps keep them from spraying your planting.

If you are concerned about the success of a new seeding, walk the site and look for seedlings. If the site was drill-seeded, look for anything growing in rows. Seedlings are quite small the first growing season. Get down on your hands and knees, and carefully inspect the site. If the success of the seeding is being challenged, hire a botanist to look for seedlings. It's worth the expense to avoid unnecessary reseeding. Otherwise, be patient for the first 2 years. If there are obvious bare areas, go back and reseed them. Normally these areas had standing water or lots of erosion.

Herbicides made to kill broadleaf weeds will also kill wildflowers and thus have to be applied carefully. Use a low-volume backpack sprayer; Birchmeier and Solo are good brands. A 4-foot wand reaches right down to the target plant, minimizing damage to surrounding plants without much bending over for the applicator (see fig. 6-2). In native plantings, use an herbicide that works on tough weeds and does not have a lot of residual effect, such as clopyralid. Follow the rates on the label for mixing the spray.

As with any prairie restoration, prescribed fire is a valuable tool for establishment and maintenance of roadside prairie plantings. To reduce thatch build-up, try to burn every 5 to 6 years. More frequent burning might be necessary to control small, woody plants. With a plowed field on one side and a gravel road on the other serving as built-in firebreaks, roadsides can be some of the easiest places to burn. And fire is one tool that readily conforms to roadside contours. Heavy traffic and utilities in the ditch make burning more complicated, requiring a larger, more experienced crew and more equipment. Caution signs must be posted, and personnel must know how to use radios and divert traffic in a way that's safe for them and for motorists. Always burn so smoke is carried away from the road. Monitor weather forecasts to anticipate changes in wind direction.

It's not surprising that, due to liability concerns, more and more plantings are not getting burned. Haying and even grazing would be the next best options but are not practical in most rights-of-way. A boom-mounted flail mower is sometimes used to cut down standing vegetation and let in some sunlight. However, this is not a perfect substitute for fire. It's important to manage thatch build-up as well. Some plantings that appear to be getting along okay without prescribed burn management will deteriorate in the long run. To protect your investment and experience the full glory of native plantings, the use of fire is strongly recommended.

To provide safe conditions for the traveling public, roadside managers routinely cut and remove woody species. Cut stumps are treated with herbicides such as Pathway and Garlon 3A. Pathfinder II is applied as a basal bark treatment; Krenite is commonly used in the fall (see table 4-2). Always follow label instructions. These practices are all beneficial to native plantings, because they eliminate trees and shrubs that otherwise produce too much shade for most prairie species.

Summary

- Two of the greatest advantages to using native vegetation are controlling soil erosion and reducing storm water runoff.
- Steep slopes require erosion control measures such as heavier seeding rates, mulch, cover crops, and, often, erosion control blankets.
- Avoid reconstructing prairie on sites previously infested with invasive species such as reed canary grass, crown vetch, and bird's-foot trefoil.
- Conduct a thorough inventory of roadsides to document the location of prairie remnants and prevent their destruction in the future.
- To create the most self-sustaining roadside vegetation, include native species adapted to a wide range of growing conditions as well as species that serve different functions within the planting.
- When roadside projects must be seeded during the summer, try to plant a temporary cover crop at that time; then return to plant the native mix at a more favorable time such as late fall or the following spring.
- The native grass drill is the preferred seeding method on level rights-of-way. Special care must be taken not to bury the seed too deep.
- Regardless of seeding method, packing the soil after planting improves germination and seedling survival.
- To maintain sightlines for motorist safety, plant turf grass or other mowable species at intersections and driveways.
- Mow plantings at least three times the first growing season to reduce competition from weeds and nurse crops.
- Native plantings benefit greatly if prescribed burns can be conducted every 5 or 6 years. In lieu of burning, haying or some other management practice must be substituted to reduce thatch build-up and destroy woody plants.

Small Prairie Plantings

KIRK HENDERSON

Overview

Small prairie plantings are used as backyard habitats, outdoor classrooms, community entryways, and low-input landscaping (reduced mowing, watering, fertilizing, and pesticide use) around homes and businesses. These plantings can range in appearance from natural-looking attempts to reconstruct a piece of prairie to more formally designed prairie gardens. Placed in high-visibility locations and viewed up close on a daily basis, these plantings are not necessarily a low-maintenance proposition. When 30 species are planted in a small area, a lot of labor can be invested trying to maintain balance. Ultimate satisfaction may depend on loosening expectations and learning to appreciate a little chaos. The rewards for establishing such personal refugia include increased familiarity with native flora and the joy of working with such a visually dynamic system.

The Planting Site

Choose a location that receives full sunlight for most of the day. To incorporate the planting into the surrounding landscape, take advantage of what the property has to offer. Perhaps there is a side yard, an area between two structures, or some otherwise well-defined parcel that can be completely or mostly filled by the planting. Use the planting to cover an entire slope or follow the length of a drive or swale. Interesting things can be done with small areas. And even a small planting is an opportunity to do less mowing.

What you don't want is a small rectangle of prairie all by itself in the middle of a turf grass lawn. The neighbors will wonder when you are going to come back and finish mowing. When working with a large campus or other expanse of lawn, fill the area as much as possible. Generally, the larger the planting, the better it will look. If the entire space is not available, shape the planting around

Fig. 13-1. The mowed border parallels the street and curves back to the building, allowing for a traditional mowed lawn between the building and the street. Photo by Kirk Henderson.

all or some of its borders, or use the planting as a transition zone between a mowed area and taller trees and shrubs (fig. 13-1).

If you lack a landscape architect's eye for shaping the planting to fit the site, a garden hose might help. As a flexible line on the ground, the hose can be moved around until a desirable size and shape are achieved. Consider leaving the hose buried in the grass as a permanent reminder of where to stop mowing the lawn and where to start managing the prairie. If the planting is to be managed with fire, maintain at least a 6-foot-wide mowed path between the planting and any woody vegetation or buildings.

Compared to traditional landscaping most prairie plantings look somewhat wild. Unless the intent is to test the local weed ordinance, think twice before establishing a front yard prairie in a very residential neighborhood. Chances are good that someone will be offended and a squabble may ensue. At the very least, front yard plantings need a wide, mowed border to set them back from the street and a buffer zone to protect them from a neighbor's herbicides.

To make prairie plantings more acceptable to neighbors, there are a few measures worth trying. Maintain a cleanly mowed border around the planting as an indication of your management and care. Use lawn timbers or a rock border to help define the area. A split rail fence provides a physical sense of containment, even with just a couple of sections positioned at the corners of your prairie. Birdhouses and birdbaths placed in the planting may help neighbors connect with the habitat aspect. And if all else fails, there's always direct communication. Go talk to your neighbors and explain what it's all about.

Seedbed Preparation

Don't worry about soil quality. Native plants do just fine in poor soil. Extremely sandy or poorly drained soils will affect your selection of plant species, but they are no cause for hauling in topsoil or applying fertilizers. In fact, many people prefer poor soil prairies, finding them more attractive and more approachable. The plants don't grow quite as tall or get quite as thick. And poor soil probably means less weedy competition during establishment.

When it's time to prepare the seedbed, leave behind all preconceived notions related to conventional farming or gardening. Unless you are working in unusually compacted soil, there is no advantage to be gained by turning over or deeply tilling the soil. Such cultivation only stirs up more weed seed. The ideal seedbed is firm and smooth.

On bare-soil sites, seedbed preparation amounts to using a rake to scratch and slightly loosen the soil surface just before seeding. When you are not working with bare soil, the most important step is to sufficiently destroy any existing vegetation prior to planting. On sparsely vegetated sites, one herbicide application will be sufficient. Apply a glyphosate herbicide when existing vegetation is green and growing and no taller than 12 inches. Always follow label instructions when using herbicides.

Yard prairies are often installed into thick turf grass. Turf grasses can be quite persistent and need to be dealt with harshly right from the beginning. The time and effort this requires will pay off in the long run. Plan to make two glyphosate herbicide applications. After the first application, wait 2 weeks; then mow off the dead vegetation. Wait another 2 weeks for any regrowth to occur, and apply the herbicide a second time. Start this process in late April or early May so your site is ready for seeding sometime in June.

Prairies can also be seeded in the fall. These are called dormant seedings. Begin applying the herbicide in September to be ready to plant right after the first frost. The seed will lie on the ground through the winter and germinate in spring. This natural cold treatment actually improves germination for some native forb species.

If herbicides are not an option, the site can be prepared by smothering existing vegetation with heavy black plastic. Leave the plastic in place, cooking the vegetation underneath, for at least a month—the longer the better. Another method of seedbed preparation is mechanical removal of the sod. Motorized sod cutters, available at rental stores, are fairly easy to operate.

Once the old vegetation has been destroyed, look the site over to decide whether the dead vegetation is so thick that it might prevent too much seed from reaching the soil. If that appears to be the case, conduct a controlled burn to get most of the dead vegetation out of the way (fig. 13-2). If burning is not an option, mow the site very short. After burning or mowing, rake the site clean and scratch up the soil a little in the process.

Consult local ordinances regarding the use of fire, and notify neighbors and the appropriate authorities before burning. Never burn alone. Have plenty of water on hand. And enlist the help of persons experienced in the use of prescribed fire.

Seed

Buy local seed. When ecologists recommend using local seed, they don't mean that you should purchase your seed from the local garden center. An ecologist's "local" refers to the seed's genetic origin. They want everyone to plant seed descended from the tallgrass prairie that was once in the vicinity of the planting or that derives from areas as close by as possible. In theory, that seed will be best adapted to local growing conditions. Plus there is a nice resonance to the idea of restoring what was actually once there and possibly providing local seed for others in the future.

When it comes to seed, opinions vary as to how local is local. Short of hand-collecting seed from nearby prairie remnants (great idea, but get permission), buying seed from a nursery or grower that specializes in native seed is the best way to live up to this ideal. Ideally, there is a native seed nursery within 200 miles of your planting. Always request local seed even if the nursery is very close by.

Don't be tempted by inexpensive, generic wildflower mixes. Packaged for nationwide distribution, these invariably contain inappropriate species—native somewhere else or not wildflowers at all. The planting will be more authentic and better adapted if it includes only those species native to your area. To minimize regrets, buy seed from a grower or nursery in your area that specializes in native seed. To search native seed sources online, type in key words such as "prairie," "native," "seed," "nursery," and the name of your state.

Use plenty of seed. A prairie planting is not a vegetable garden, where bare soil is maintained between rows of plants. Part of the strategy for suppressing

Fig. 13-2. Conducting a prescribed burn helps get dead plant material out of the way prior to seeding. Photo by Kirk Henderson.

weeds and reducing maintenance is to have all available space occupied by native plants right away. To that end, plant at least 80 seeds per square foot. Very roughly, that would be 8 ounces of pure live seed for 1,000 square feet. On large, level planting sites, that would be a heavy and unnecessarily expensive rate of seeding. On small plantings, it amounts to affordable insurance.

When putting together a seeding mix, emphasize diversity. Include many different native species to create a planting that is ecologically sound and visually interesting. Working with small areas allows you to load up on forbs without adding too dearly to the cost of the planting. Take advantage of this fact by experimenting with lots of different forbs. It's not totally predictable which species will thrive on a particular site, nor is it easy to predict how individual species will get along.

Grasses are equally important to the look of the planting, so do not skimp on this important component. Native grasses come in handsome shapes and sizes and are a source of color throughout the season. A good grass matrix provides spacing and a sense of order. Grasses also provide competition and support for forbs that might otherwise get too rangy and flop over. Put together a seeding mix that is pretty evenly divided between forbs and grasses.

Fig. 13-3. The prairie dropseed plants in the foreground are at full height and are as attractive as any ornamental grass. Photo by Kirk Henderson.

In small prairies where diversity is the goal, go easy on the taller grasses. For big bluestem, use no more than one-fourth ounce of pure live seed per 1,000 square feet. The same goes for Indian grass. Seed little bluestem and side-oats grama each at an ounce of pure live seed per 1,000 square feet. And try to work in some prairie dropseed, a beautiful, shorter-growing native grass, best started from live plants (fig. 13-3).

If soil conditions tend decidedly to one extreme or the other, that is to say, either very wet and heavy or very sandy and well drained, tailor the species mix to fit those conditions. And if the site is fairly shaded, purchase a mix designed for savanna or open woods. Otherwise, a diverse upland prairie mix (30 species or more) will result in the successful establishment of plenty of species.

Taller and More Aggressive Native Species

Some plants become overly dominant in small areas, threatening the visibility or even the survival of other species. Smaller plantings will retain their original diversity much better over time if the taller and more aggressive species are kept to a minimum. And smaller plantings tend to look better if they don't contain so many of the really tall prairie species. If you want to include a specimen of any of these plants, consider putting in a couple of live plants instead of scattering their seed.

Some of the taller and more aggressive native species that should be kept to a minimum or omitted altogether include cup plant, New England aster, ox-eye sunflower, saw-tooth sunflower, Maximilian sunflower, and even big bluestem.

Black-eyed Susan, wild bergamot, and gray-headed coneflower may come on strong at first but shouldn't provide too much competition in the long run. Still, use them sparingly so they don't crowd something out early on.

On the other hand, this is the region of the tallgrass prairie. Unless you use a really limited seeding mix, the planting will include many species that, by late summer, are 4 feet tall or taller. Native plantings can exhibit sort of a split personality, resembling more of a traditional flowerbed early in the year and a small piece of wilderness by late in the growing season (figs. 13-4 and 13-5). Be prepared to embrace this reality.

Many native plants are not as showy as typical garden-variety plants. These plants with inconspicuous flowers are no less important in a native plant community. The prairie flora includes many species whose physical appeal derives from a distinctive shape of leaf and stem. Others have no appeal by traditional garden standards.

Native-seed dealers have catalogs and websites listing the species they carry. Some of these have photographs. Catalogs can be a good source of information about each species' habitat preference, growing height, and blooming time. Some dealers offer seed that is premixed for various situations. These mixes can be helpful. However, often no information is provided about how much of any one species the mix includes.

Scatter the seed by hand, making sure that plenty of seed is applied all the way out to the very margins of the planting. These edges are where the planting is most vulnerable to encroachment by surrounding nonnative species. When working with small quantities of seed, it's easy to run out before the entire area is covered. For better seed distribution, stir the seed into a bucket or pan that is half full of clean sand. Divide the sand and seed mixture into fourths, and seed one-fourth of the planting area at a time.

Lightly rake the seed into the soil to a depth of no more than one-quarter inch. After seeding, pack the soil tightly with a roller, or drive over it with a tractor. At the very least, walk the seed into the soil. Packing improves germination and seedling survival. Don't worry about packing too tightly. The seed will not be damaged.

A naturally patchy appearance can be achieved by planting concentrations

Fig. 13-4. In the spring when plants are small, tallgrass prairie plantings can resemble a traditional garden. Photo by Kirk Henderson.

Fig. 13-5. By late summer, the same planting resembles a small piece of wilderness. Photo by Kirk Henderson.

of individual species or small sets of species in selected areas. These concentrations can be isolated or overlapping. Place them randomly, or follow a plan. This mosaic seeding method can also be employed to position taller species at the back of the planting or down the middle, with shorter species in front or around the edges.

Consider putting together a matrix of less competitive species such as black-eyed Susan, partridge pea, round-headed bush clover, tall dropseed, little bluestem, and side-oats grama and scattering it uniformly throughout the planting. This method insures establishment of at least something over the entire planting area. Then seed the species concentrations over the top of that.

Live Plants

Small plantings can be greatly enhanced with the addition of live plants, even just a few. Spend most of your live plant budget on species that do not establish readily from seed. The following list includes plants that aren't as likely to establish from seed as well as some that bloom in early spring, when it's harder to get color into the planting. The list includes unusual plants and popular ones that will add to the quality and visual appeal of the planting. These plants are not native to all locations. Seek out your own regional gems. Suggested live plants include wild garlic, leadplant, Canada anemone, butterfly milkweed, silky aster, cream false indigo, New Jersey tea, shooting star, pale purple coneflower, rattlesnake master, prairie smoke, alumroot, yellow stargrass, round-headed bush clover, prairie blazing star, Michigan lily, great blue lobelia, wild quinine, prairie phlox, prairie rose, little bluestem, blue-eyed grass, and prairie dropseed. Also, consider establishing big bluestem and Indian grass strictly from live plants as a way to control the number and location of these taller grasses.

When ordering live plants, purchase the largest plants that your budget allows, especially if irrigation of the site is not possible. Tiny plants with little root mass are quick to shrivel and die during dry spells. Install the live plants right after the site has been seeded.

The smaller the planting, the easier it is to establish the whole thing from live plants. Live plants are more expensive than seed, and there's more labor involved. But live plants yield faster results and provide more control over the planting's eventual appearance. Live plants are recommended where a much more managed appearance is desired.

Woven weed barrier, also called landscape fabric, adds cost and labor to the project. It does provide effective weed control when you are working strictly with live plants. Weed barrier is especially useful for maintaining spacing between plants and when working in formal settings. Weed barrier can be removed after a couple of years, once plants are well established.

Before installing weed barrier over turf grass or other existing vegetation, prepare the site with a glyphosate herbicide as described above, waiting 2 weeks after spraying before installing plants. It's possible to forgo the herbicide treatment if the weed barrier is put down far enough in advance of planting, in March or the previous fall. Purchase weed barrier of good quality, made of heavier material and with good UV stabilization.

Stretch the fabric tight. Overlap adjoining strips by 3 inches. Pin the weed barrier securely in place with 6-inch metal staples placed every 2 or 3 feet. Wait several days for the puffiness to go away as the weed barrier and underlying plant material flatten to the soil. This will make it easier to install the plants. The weed barrier makes a nice artist's canvas for onsite designing or transferring a design onto the site. Marking the location for individual plants or groups of plants is a good idea, especially when a labor crew does the installing.

To install the plants, cut an x in the weed barrier just big enough to dig the hole and accommodate the plant. When using bare rootstock, dig the hole wide enough to spread the roots and as deep as the root is long. Holes for tiny plants can be made with a dibble. For larger plants, a soil auger powered by a ⅜-inch, high-voltage drill will make the job easier. The root crown should be just below the soil surface without mounding the soil. A shallow, dish-shaped depression around each plant makes watering much easier. Fill the hole and pack the soil to remove all air pockets, and generously water the plants in.

Do not expose bare roots to sun and wind during the planting process. Carry them in a bag of peat moss, or keep them in a bucket of water during planting. Fine root hairs dry out incredibly fast, harming the plant's chance of survival. Order approximately one plant for each square foot of planting space.

Covering the weed barrier with mulch is important for aesthetic purposes when spacing needs to be maintained between plants, or if it matters how the planting looks before plants mature and hide the fabric. If the planting is to be used as a teaching aid, wider spacing makes it easier to identify individual species.

Use good-quality mulch. Wood chips from a garden center last longer and are weed-free compared to those from the local landfill. There is no way to completely weed-proof the planting, since weed seed blown into clean mulch can germinate and grow, too. Be careful not to smother the seedlings by piling mulch too high or too close.

As an alternative to expensive weed barrier and wood mulch, put down a layer of newspaper and cover it with leaves or grass clippings. These materials will last long enough to help plants get established and then decompose after a couple of years, allowing for more natural reproduction.

Maintenance

During the first growing season, prairie plants invest their energy sending down roots; they exhibit very little above-ground growth (fig. 13-6). During the first year, these tiny seedlings are at risk of being shaded out by tall weeds. Prairie seedlings can be difficult to identify also. This increases their chances of being disturbed or accidentally removed during any hand-weeding. The safest way to control weeds is to mow the planting to a height of 4 to 6 inches every 3 weeks throughout the first growing season. Weeds squeezing through the holes made in weed barrier should be snipped off instead of pulled. If irrigation is possible, turn on the sprinkler hose with the goal of providing an inch of water every week.

Fig. 13-6. It's difficult to anticipate what weed species will be released when existing vegetation is removed. Shown in late July, this planting is dominated by crab grass during its first growing season. Photo by Kirk Henderson.

Fig. 13-7. The second growing season brings on an explosion of black-eyed Susan. This is the only year the planting will look like this. Photo by Kirk Henderson.

Fig. 13-8. Wild bergamot is most conspicuous, but the planting in its third growing season begins to exhibit quite a bit of diversity, with many other plants starting to show up as well. Photo by Kirk Henderson.

Fig. 13-9. By the fourth growing season, biomass at ground level provides enough fuel to carry the fire. Fire helps eliminate a heavy build-up of thatch. Photo by Kirk Henderson.

During the second year (fig. 13-7), fill bare spaces with additional seed or with transplants. By the third or fourth growing season, some of the established plants can be divided and used to fill bare areas and improve distribution. Random weeds should be pulled or snipped off with hand trimmers before they produce seed or get too well established.

By the third year, even small plantings produce a tremendous amount of above-ground biomass (fig. 13-8). Fire is the best way to remove this material and prevent the build-up of a heavy layer of thatch. For general maintenance, burn the planting every 3 to 5 years (fig. 13-9).

Consideration for neighbors and local burn ordinances can make prescribed burning difficult or impossible. In this case, mowing is the best substitute for fire, as long as the cuttings are raked up and removed. It takes a heavy-duty mower to get through a mature planting.

Alternatives to mowing include cutting with grass clippers, hedge trimmers, or even a steel-bladed weed eater. Cutting and removing is a lot of work but makes it easier to enjoy emerging plants in spring. Do not get overzealous with spring clean-up, because butterfly eggs and larvae overwinter in the leaf litter.

For plantings in urban locations, it can be a good public relations move to mow the planting at the end of every fall so dead plant material does not attract negative attention through the winter. On the other hand, many of us enjoy seeing the dried plants silhouetted against the snow. And it's fun to watch juncos shake the stems of plants to eat the seeds off the snow.

To get to know the species and keep track of where they were planted, individual plants can be labeled with small markers stuck in the ground. Tags made for labeling plants can be purchased through catalogs and garden centers. Signs placed in plantings help people understand and appreciate what's going on. A "Prairie Restoration" sign tells neighbors that the tall stuff is there on purpose and is not the result of neglect. For young plantings, a "Prairie in Progress" sign is used as a request for patience.

Variations in site conditions and species composition make every small prairie planting an experiment with somewhat unpredictable results. The commitment of time and energy is as ongoing as it is in any garden project. As the planting develops, so does your relationship with the individual plants and the unique habitat they create.

Summary

- Yard prairies reduce the need for mowing, watering, fertilizers, and pesticides but are not necessarily a low-maintenance landscape.
- Native species do just fine in poor soil; many people prefer the look of poor-soil prairies.
- Turf grass or other existing vegetation must be thoroughly destroyed prior to seeding.
- Deep tillage should be avoided to prevent bringing more weed seed to the light of day.
- Smaller plantings will retain their original diversity much better over time if taller and more aggressive species are kept to a minimum.
- The seeding mix should be evenly divided between native forbs and grasses when the goal is to create a natural-looking prairie planting.
- Live plants can be used to enhance seeded prairies, or they may be used exclusively to achieve a more formal appearance.
- Apply seed at a heavy rate, at least 80 seeds per square foot, all the way to the edges of the planting.
- Mow the planting every 3 weeks throughout the first growing season.
- Some native species are aggressive and may require regular weeding.
- Burn or mechanically remove above-ground growth every 3 to 5 years.

Native Seed Production

Seed Harvesting

GREG HOUSEAL

Overview

Seed of many native species is now commercially available for prairie reconstructions, large or small. Yet many people have an interest in growing and collecting native species for butterfly gardens, backyard and schoolyard wildlife habitats, and prairie restorations. Native seed may be harvested as single or mixed species from remnant or reconstructed prairies or from seed nursery production fields. Harvesting techniques range from hand-collecting to hand-operated machines to combines. General guidelines for harvesting from remnant prairies and capturing genetic diversity are presented in this chapter. Awareness of the indications of seed ripening and environmental factors affecting seed maturity will aid proper timing of native seed harvest.

Harvesting from Remnant Sites

Be mindful of legal and ethical considerations when collecting seed from prairie remnants. Federal and state endangered and threatened species cannot be collected without proper permits. Also, repeated annual harvesting of the same remnant area should be avoided. Manipulation of a remnant prairie to maximize seed production — such as repeated annual burns on the entire site, herbicide treatments, or fertilizing — is inappropriate. A remnant prairie is a diverse, biotic community (both above and below ground) of microbes, fungi, plants, and animals (vertebrate and invertebrate) interacting in complex relationships. Any management applied indiscriminately and repeatedly will be detrimental to some of these associations. Even burning should be limited to only a portion of a remnant in any given year, and each portion should be burned on rotation and at different times of the year at varying intervals.

A commonly expressed rule is "take half, leave half" when harvesting seed from remnant sites. The objective is to avoid overcollecting any species from a site. Overcollecting is a real possibility on small, high-quality remnants with public access. The negative impacts of overcollecting are not limited to the loss of seed itself; they also include the trampling of vegetation and the introduction of exotic or invasive plants brought in on clothing or machinery. However, most of the seed produced in any given year will not have the opportunity to germinate, grow, and establish itself as a mature plant in the community. Seed is also lost to predation, disease, and decay. Likewise, most prairie species are perennial, meaning that their roots survive over winter to regrow shoots the next spring, so an annual seed crop is not essential to the perpetuation of the population. Exceptions to this are annual, biennial, and short-lived perennial species; rare and uncommon species; or common species poorly represented in a remnant.

Of course, it's essential to ask permission of the landowner or appropriate land management agency before collecting seed on private property. Likewise, removal of any plant or plant part from preserves, natural areas, and parks may be restricted. Check with the proper agency before harvesting seed in these areas. Harvesting from roadsides may also be restricted in some states and counties. Many counties in Iowa, for example, are planting native prairie in roadside rights-of-way, and some may harvest seed from those rights-of-way in subsequent years. Ask permission from the county roadside manager or engineer or state Department of Transportation before harvesting from roadsides.

Finally, any mechanical harvesting occurring in remnant sites should include a careful inspection and cleaning of equipment prior to use, including vehicles, to avoid introducing exotic or invasive species that may lead to the degradation of the remnant or create long-term management issues.

Collecting Seed for Genetic Diversity

An important restoration goal should be to capture genetic diversity from remnant populations. Here are some rules of thumb to guide your efforts. Of course, be reasonably sure that the site is a remnant—never plowed, not planted. Then, keep in mind two important ideas. First, attempt to *collect roughly equal amounts of seed from several individuals* in the population. Second, generally speaking, near neighbors are more closely related genetically than distant individuals, so it is important to *collect seed from throughout the population.*

Collect seed from at least 20 to 30 well-dispersed individual plants within a population, if possible. Randomize the process to avoid intentionally selecting plants based on size, color, vigor, or any other trait. The point is to capture genetic diversity, not novelty. To sample large populations, walk transects and collect seed perhaps every 10 paces. Collect roughly equal amounts of material (seed or seed head) from each plant you encounter. If collecting from multiple sites, attempt to equalize the contribution of seed from each site, particularly if collecting seed as foundation stock for nursery production to generate seed for other reconstructions. Populations grown and regrown in a production field can become adapted to site conditions and nursery management practices. Therefore, it is important to save seed from the original collections or from earlier generations for replanting production fields. Periodically reinfuse seed stock with remnant-collected seed when possible.

General Indications of Seed Ripening

Seed ripening and timing of harvest vary by species, source of parent material, and environmental conditions (table 14-1). Most species ripen gradually, so not all seed will be at the same stage of maturity at any given time. Seed maturity usually progresses from top to bottom of the seed head in grasses and many prairie forbs. Mature seed may be quickly dispersed either by gravity, wind, water, or animals, so it's important not to delay harvesting. Immature seed stores poorly, losing viability more quickly than mature seed.

In terms of the plants' environment, cold, moist conditions will tend to delay seed maturity, while hot, dry conditions may hasten it. Latitude will also affect ripening, since many plants flower and set seed in response to photoperiod. Flowering and seed set may be delayed if plants are grown north of their origin or hastened if they are moved south.

Among grass species, cool-season grasses begin growth early in the growing season and consequently ripen earlier compared to warm-season grasses. There are roughly four stages of grass seed maturity: milk, soft dough, medium dough, and hard dough. Firm thumbnail pressure on the caryopsis, or germ, will help determine maturity. Grasses should be harvested at the hard-dough stage, when the caryopsis resists firm thumbnail pressure. Many grasses do not hold seed long after maturity. Test ripeness by firmly striking the seed head against your palm; if some shattering occurs, the seed is ready to harvest. If it shatters with only gentle striking, harvest immediately. Since most grasses ripen from the top down, some shattering of the tops of the seed heads may have already occurred.

In forb species, the seed head itself or the stalk immediately below the seed head will begin to appear dry and discolored as the seed ripens. Notable exceptions are the spiderworts, members of the dayflower family. These species will drop seed from individual flower heads as they ripen even while the bracts

Fig. 14-1. Species with dispersal apparatus such as a pappus, like this rough blazing star, will appear dry and fluffy at maturity. Photo by Greg Houseal.

remain green and other flowers in the same cluster are in bud or blooming. Species with dispersal apparatus such as pappus (parachutes) will appear dry and fluffy at maturity (fig. 14-1). Some species forcefully eject seed at maturity—phlox and violets, for example—and must be checked and harvested daily or bagged loosely with a tightly woven mesh or cloth bag or nylon hosiery. Refer to table 14-1 for seed-ripening times for selected tallgrass species.

Harvesting Techniques

Harvesting techniques range from hand-collecting to hand-operated machines to combines. Any species can be collected by hand, but hand-collecting is particularly useful when you are collecting seed of native species which occur on specific sites that may be inaccessible by machines, grow very low or very high or ripen early or late, occur as uncommon or patchy species in native prairie, or have explosive seed-dispersal mechanisms.

Hand-harvesting is also a good volunteer group activity for introducing people to prairies and a good way to collect very local seed for nearby prairie restorations. Hand-harvesting is time- and labor-intensive and perhaps not practical for large projects. Efficiency can be improved by keeping both hands free to

Table 14-1. Harvest Periods for Tallgrass Species in Iowa

							Optimal Forb Early Season			Optimal Forb Mid Season							Optimal Forb Late Season						
			May			June			July			August			September			October			November		
Species	Physiognomy		1-10	10-20	20-30	1-10	10-20	20-30	1-10	10-20	20-30	1-10	10-20	20-30	1-10	10-20	20-30	1-10	10-20	20-30	1-10	10-20	20-30
Native forbs																							
Marsh marigold	Perennial	Forb																					
Prairie smoke	Perennial	Forb																					
Pussytoes	Perennial	Forb																					
Indian paintbrush	Annual	Forb—hemiparasite																					
False dandelion	Perennial	Forb																					
Swamp saxifrage	Perennial	Forb																					
Golden ragwort	Perennial	Forb																					
Blue-eyed grass	Perennial	Forb																					
Yellow stargrass	Perennial	Forb																					
Lousewort	Perennial	Forb—hemiparasite																					
Prairie violet	Perennial	Forb																					
Spring cress	Perennial	Forb																					
Wild garlic	Perennial	Forb																					
Hoary puccoon	Perennial	Forb																					

							Optimal Forb Early Season			Optimal Forb Mid Season									Optimal Forb Late Season				
			May			June			July			August			September			October			November		
Species	Physiognomy		1-10	10-20	20-30	1-10	10-20	20-30	1-10	10-20	20-30	1-10	10-20	20-30	1-10	10-20	20-30	1-10	10-20	20-30	1-10	10-20	20-30
Marsh phlox	Perennial	Forb																					
Prairie phlox	Perennial	Forb																					
Canada anemone	Perennial	Forb																					
Self heal	Perennial	Forb																					
White camass	Perennial	Forb																					
Shooting star	Perennial	Forb																					
Marsh bellflower	Perennial	Forb																					
Violet wood sorrel	Perennial	Forb																					
Bastard toadflax	Perennial	Forb—hemiparasite																					
Veiny pea	Perennial	Legume																					
Woundwort	Perennial	Forb																					
American vetch	Annual	Forb																					
Black-eyed Susan	Perennial	Forb																					
Marsh vetchling	Perennial	Legume																					
Spiked lobelia	Perennial	Forb																					
Blue flag	Perennial	Forb																					
Milk vetch	Perennial	Legume																					
Pasture rose		Shrub																					

							Optimal Forb Early Season			Optimal Forb Mid Season							Optimal Forb Late Season						
			May			June			July			August			September			October			November		
Species	Physiognomy		1-10	10-20	20-30	1-10	10-20	20-30	1-10	10-20	20-30	1-10	10-20	20-30	1-10	10-20	20-30	1-10	10-20	20-30	1-10	10-20	20-30
Illinois tick trefoil	Perennial	Legume																					
Cowbane	Perennial	Forb																					
Ironweed	Perennial	Forb																					
Culver's root	Perennial	Forb																					
Showy tick trefoil	Perennial	Legume																					
Purple prairie clover	Perennial	Legume																					
Ox-eye sunflower	Perennial	Forb																					
Narrow-leaved loosestrife	Perennial	Forb																					
Winged loosestrife	Perennial	Forb																					
New Jersey tea		Shrub																					
Joe Pye weed	Perennial	Forb																					
Purple meadow-rue	Perennial	Forb																					
Hairy mountain mint	Perennial	Forb																					
Cardinal flower	Perennial	Forb																					
Prairie lily	Perennial	Forb																					
Golden alexanders	Perennial	Forb																					
Compass plant	Perennial	Forb																					

							Optimal Forb Early Season			Optimal Forb Mid Season						Optimal Forb Late Season						
			May			June			July			August			September			October			November		
Species	Physiognomy		1-10	10-20	20-30	1-10	10-20	20-30	1-10	10-20	20-30	1-10	10-20	20-30	1-10	10-20	20-30	1-10	10-20	20-30	1-10	10-20	20-30
Rattlesnake master	Perennial	Forb																					
Nodding wild onion	Perennial	Forb																					
Butterfly milkweed	Perennial	Forb																					
Great blue lobelia	Perennial	Forb																					
Rosinweed	Perennial	Forb																					
Slender mountain mint	Perennial	Forb																					
Fringed loosestrife	Perennial	Forb																					
Round-headed bush clover	Perennial	Legume																					
Swamp milkweed	Perennial	Forb																					
Pale purple coneflower	Perennial	Forb																					
White prairie clover	Perennial	Legume																					
Leadplant		Shrub																					
Fragrant coneflower	Perennial	Forb																					
Common mountain mint	Perennial	Forb																					

							Optimal Forb Early Season			Optimal Forb Mid Season									Optimal Forb Late Season				
			May			June			July			August			September			October			November		
Species	Physiognomy		1-10	10-20	20-30	1-10	10-20	20-30	1-10	10-20	20-30	1-10	10-20	20-30	1-10	10-20	20-30	1-10	10-20	20-30	1-10	10-20	20-30
Prairie coreopsis	Perennial	Forb																					
Thimbleweed	Perennial	Forb																					
Gray-headed coneflower	Perennial	Forb																					
Swamp lousewort	Perennial	Forb																					
Boneset	Perennial	Forb																					
Tall cinquefoil	Perennial	Forb																					
Stiff goldenrod	Perennial	Forb																					
Michigan lily	Perennial	Forb																					
Prairie blazing star	Perennial	Forb																					
Cream false indigo	Perennial	Legume																					
Silky aster	Perennial	Forb																					
Sky-blue aster	Perennial	Forb																					
Smooth blue aster	Perennial	Forb																					
Wild bergamot	Perennial	Forb																					
White wild indigo	Perennial	Legume																					
Hoary vervain	Perennial	Forb																					
New England aster	Perennial	Forb																					
Rattlesnake-root	Perennial	Forb																					
Riddell's goldenrod	Perennial	forb																					

							Optimal Forb Early Season			Optimal Forb Mid Season							Optimal Forb Late Season					
			May			June			July			August			September			October			November		
Species	Physiognomy		1-10	10-20	20-30	1-10	10-20	20-30	1-10	10-20	20-30	1-10	10-20	20-30	1-10	10-20	20-30	1-10	10-20	20-30	1-10	10-20	20-30
White sage	Perennial	Forb																					
Willowleaf aster	Perennial	Forb																					
Wild quinine	Perennial	Forb																					
Rough blazing star	Perennial	Forb																					
Virginia anemone	Perennial	Forb																					
Tall coreopsis	Perennial	Forb																					
Missouri goldenrod	Perennial	Forb																					
Old field goldenrod	Perennial	Forb																					
Showy goldenrod	Perennial	Forb																					
Grass-leaved goldenrod	Perennial	Forb																					
Bottle gentian	Perennial	Forb																					
TOTAL FORB SPECIES POTENTIALLY RIPE					2	3	9	9	8	8	12	4	4	8	17	38	50	54	51	37	20	4	1

								Optimal Cool Season									Optimal Warm Season						
Species	Physiognomy		May			June			July			August			September			October			November		
			1-10	10-20	20-30	1-10	10-20	20-30	1-10	10-20	20-30	1-10	10-20	20-30	1-10	10-20	20-30	1-10	10-20	20-30	1-10	10-20	20-30
Grasses and sedges																							
Sweet grass	Perennial	Grass																					
Yellow fox sedge	Perennial	Sedge																					
Prairie star sedge	Perennial	Sedge																					
Hummock sedge	Perennial	Sedge																					
Mead's sedge	Perennial	Sedge																					
Porcupine grass	Perennial	Grass																					
Prairie wedgegrass	Perennial	Grass																					
Bicknell's sedge	Perennial	Sedge																					
Shortbeak sedge	Perennial	Sedge																					
Heavy sedge	Perennial	Sedge																					
Troublesome sedge	Perennial	Sedge																					
Brown fox sedge	Perennial	Sedge																					
Woolly sedge	Perennial	Sedge																					
Bluejoint grass	Perennial	Grass																					
June grass	Perennial	Grass																					
Bebb's sedge	Perennial	Sedge																					
Fowl manna grass	Perennial	Grass																					
Slender wheatgrass	Perennial	Grass																					

								Optimal Cool Season								Optimal Warm Season						
			May			June			July			August			September			October			November		
Species	Physiognomy		1-10	10-20	20-30	1-10	10-20	20-30	1-10	10-20	20-30	1-10	10-20	20-30	1-10	10-20	20-30	1-10	10-20	20-30	1-10	10-20	20-30
Side-oats grama	Perennial	Grass																					
Canada wild rye	Perennial	Grass																					
Dark green bulrush	Perennial	Sedge																					
Little bluestem	Perennial	Grass																					
Prairie cord grass	Perennial	Grass																					
Switchgrass	Perennial	Grass																					
Indian grass	Perennial	Grass																					
Prairie dropseed	Perennial	Grass																					
Big bluestem	Perennial	Grass																					
Tall dropseed	Perennial	Grass																					
Upland wild timothy	Perennial	Grass																					
Virginia wild rye	Perennial	Grass																					
Woodland reedgrass	Perennial	Grass																					
TOTAL GRAMINOID SPECIES POTENTIALLY RIPE								8	10	10	3	1	1		1	9	10	12	7	4	0	0	

Note: Shaded areas represent optimum collection periods, when the most species are likely to be in fruit.

Sources: Iowa Ecotype Project; Iowa NRCS staff biologist Jennifer Anderson-Cruz; Packard and Mutel 2005.

harvest by fastening collection bags and containers around your waist. Leather gloves and good-quality scissors or shears are a must for effective seed collecting. Unbreakable plastic combs are inexpensive and efficient tools for stripping grass seed. Choose brightly colored tools that will be easy to spot if you drop or misplace them while collecting. Use breathable bags (cloth or paper) for collecting that will allow moisture to escape. Even seemingly dry seed and seed heads retain enough moisture when first collected to cause mildew or rot if left in plastic bags. Use care not to leave collected material in closed vehicles that may heat up in the sun.

Seed can be stripped by hand from many species such as blazing stars, asters, and grasses. In species with seed in salt-shaker pods, such as shooting star, great St. John's wort, larkspur, and columbine, try tipping the pod into an open container. This will minimize the need to clean the seed later. If seed is held tightly in the seed head, simply clip a portion of the seed head for later cleaning. Prickly seed heads like rattlesnake master or pale purple coneflower will require gloves and shears for efficient collecting.

Mechanical harvesting can be accomplished with modified combines or native seed strippers. Combines are generally more efficient at capturing seed, particularly if you are harvesting monospecific seed-production fields, but they are used for harvesting mixed species stands as well. Commercial seed strippers are available as handheld, implement pull-type, or tractor-mounted equipment. They all use a rotating brush or bristles to "strip" the seed from stems and stalks. While perhaps not as efficient as combines, strippers can be used for the successive harvest of species that ripen gradually or at different times. If leaving wildlife habitat intact over winter is a priority, seed strippers remove seed while leaving the overall structure of the vegetation intact, whereas combines cut stems off at an even height. Handheld strippers and pull-types light enough to be pulled with an ATV allow harvest of sites that are otherwise inaccessible.

Because not all seed ripens at the same time, mechanical harvesting requires you to determine when most seed is at or nearing maturity. If a species' seed shatters very easily, harvest in the early morning when humidity is high and wind is calm, since a strong wind can reduce the harvest significantly in a single afternoon. Windrowing or swathing during the medium- to hard-dough stage in grasses that otherwise shatter easily at maturity can be effective, since seed will continue to ripen for several days after cutting. For this method, however, it's important to be sure no rain is in the forecast for the next few days after cutting. Swaths can then be picked up with a combine after the material has dried at the site for a few days. Combines may require significant modifications to make them suitable for harvest of native grasses and forbs (fig. 14-2).

Fig. 14-2. Combining stiff goldenrod with a 1978 Hege 125B plot combine with a modified boot attachment on rear. Photo by Greg Houseal.

Species with fluffy seed such as asters and goldenrods should be harvested when seed is mature, but just prior to the dry-fluff stage. If the seed is dry-fluffy, the combine will become a super seed-dispersal machine! Reducing or shutting off airflow in the combine is a must for native species with lightweight seed. Plugging of shaker sieves and augers is a constant issue, particularly with fluffy seed or seed with long awns. The long, twisty awns of Canada wild rye can be a combiner's bane. Deawner bars can be installed into the cylinder surrounding the concaves to increase the threshing action of the concaves.

Keeping Records

Keeping records of where and when you collect seed provides important information about a prairie restoration. Include basic information: location (county, township, range, section, quarter section), soil type (sandy, clayey, loamy) and moisture (wet, medium, dry), slope and aspect (direction the slope faces), approximate size of the population, number of plants collected from, and the date (fig. 14-3). It's a good idea to include a sketch of the site to jog your memory about where the species occurred within the prairie.

Fig. 14-3. Sample seed-collecting tag.

Seed collection label			Date collected:	
Collector(s):				
Address:				
Contact information:				
Species collected:				
County	Township	Range	Section	Quarter section
Property owner/land management organization			Sketch of site	
Soil type:				
Slope:				
Aspect (direction slope faces):				
Approximate size of population:				
No. of individual plants collected from:				
Associated species:				

Summary

- Native seed may be harvested as single or mixed species from remnant or reconstructed prairies or from seed nursery production fields.
- "Take half, leave half" is a reminder not to overcollect from a site.
- Federal and state endangered and threatened species cannot be collected without proper permits.
- Repeated annual harvesting of a given remnant area should be avoided.
- Manipulation of a remnant prairie to maximize seed production—such as whole-site annual burns, herbicide treatments, or fertilizing—is inappropriate and unethical.
- Ask permission of the landowner before collecting seed on private property.
- Removal of any plant or plant part from preserves, natural areas, or parks is restricted, and permission must be obtained from the proper authorities.
- Any mechanical harvesting occurring in remnant sites should include a careful inspection and cleaning of equipment and vehicles prior to use to avoid the introduction of exotic or invasive species that could degrade the remnant or create long-term management problems.
- Attempt to collect roughly equal amounts of seed from several individuals in the population, and collect seed from throughout the population.

- Cold, moist conditions tend to delay seed maturity, while hot, dry conditions may hasten it.
- Latitudinal location will affect ripening, since many plants flower and set seed in response to photoperiod.
- Grasses should be harvested at the hard-dough stage, when the caryopsis resists firm thumbnail pressure.
- Generally, the seed head itself or the stalk immediately below it will begin to appear dry and discolored as the seed ripens.
- Most species ripen gradually, so not all seed will be at the same stage of maturity at any given time.
- Hand-harvesting can be an important way to collect seed of native species that otherwise are unavailable or inaccessible to machine harvest, such as low- or high-growing species, species that ripen early or late, uncommon or patchy species in native prairie, or species with explosive seed-dispersal mechanisms.
- Hand-harvesting is time- and labor-intensive and not practical for large projects.
- Commercial seed strippers are available as handheld, implement pull-type, or tractor-mounted equipment.
- Combines may require significant modifications to make them suitable for the harvest of native grasses and forbs.
- Swathing grass species that shatter easily during the medium- to hard-dough stage can be effective, since seed will ripen for several days after cutting.
- Keep records of where and when you collect seed.

15

Drying, Cleaning, and Storing Prairie Seed

GREG HOUSEAL

Overview

Postharvest processes include drying, precleaning, cleaning, and storing seed properly. If the seed is collected in bulk and immediately spread on a restoration site, little processing is necessary. If the seed is to be stored for any length of time, the next step is to properly care for the harvest. Drying, cleaning, and storage requirements for prairie seed after collecting will depend on how and which species are collected, the length of time stored, and the intended seeding method. Keeping seed of individual species separate will aid cleaning and assessing of seed quality and amount. Provisions should be made to begin drying any material stored more than a day. Knowledge, skill, and access to specialized equipment are necessary for some of the cleaning steps described, and these factors will determine the quality of the finished product. Most commercial producers of native seed use all these methods to clean seed to a very high degree of purity and germination.

Drying Seed

Drying bulk material immediately after harvest is critical for preventing mold and mildew. Drying allows some immature seed to ripen; it aids threshing of the seed out of seed heads or pods and thus helps maximize seed yield. One drying method is to spread the seed out on screening or newspaper in a cool, dry place with good air circulation. Seed can also be dried in breathable cloth collection bags if good air circulation is provided and the bags are checked for dampness and turned frequently. If using paper bags for drying seed, leave bag tops open and turn the contents once or twice daily. Take care not to pack collected bulk material into bags too tightly; keep it loose so air can circulate. Plastic bags or airtight containers can be used for storage only after the seed is properly dried and cleaned. Larger quantities of material will require special bins with screened bottoms and a source of airflow up through the material for it to dry properly (fig. 15-1). Smaller quantities can be collected in large, 100 percent cotton bags made to fit inside a 30-gallon plastic bin. Fill the bags loosely with seed heads, tie them closed, label them, and place them in the drying bins. If the material associated with the seed is very green, as is the case with spiderwort, or is damp from a recent shower, it's best to spread the material out on tarps and position several box fans overhead, turning the seed frequently with pitchforks or shovels if dealing with larger quantities. Drying may take several days to a few weeks, depending on quantity and drying conditions (fig. 15-2).

Precleaning Seed

Much of the bulk material in collected native seed is nonseed, inert floral parts, leaves, and stems. Harvested material will require some degree of precleaning to reduce bulk and improve flow for later cleaning processes. The extent of precleaning required will depend on the method of seed harvest, intended storage period, and method of planting. Threshing, debearding and deawning, and brushing are considered precleaning methods, because they are designed to prepare the seed for later cleaning processes by removing unnecessary appendages and improving seed flow. If the material was machine-combined, threshing has already been accomplished. Otherwise, you can use several low-tech hand-threshing techniques, including stomping, shaking, and screening.

THRESHING SEED

Threshing removes the seed from seed heads. Species with large, coarse seed heads that tend to hold the seed tightly can be threshed by stomping on the seed heads. This method is very effective on species of wild indigo, rattlesnake master, compass plant and rosinweed, sunflowers, black-eyed Susan and fragrant coneflower, and golden alexanders. Place about 2 inches of bulk material in the bottom of a large plastic tub, and stomp on it with waffle-sole boots. Toe kicks to the corners of the tub help break up any stubborn seed heads. Screen the stomped material through a coarse, half-inch or quarter-inch screen into a second tub. Continue in batches, returning any intact seed heads to the stomping tub. Pale purple coneflower tends to be stubborn and may require machine threshing, unless it's collected late in the season after seed heads naturally begin to break apart.

Many species have seed that shakes free of a capsule or open pod. The shake method can be effective for dried seed heads of Culver's root, cardinal flower and great blue lobelia, shooting stars, mints, and gentians. Either hold the dry

Fig. 15-1. Bagged combined seed material in forced-air drying bins. The 30-gallon bags are custom-made of 100 percent cotton with drawstring closures. Photo by Greg Houseal.

Fig.15-2. Various homemade or purchased screens and tubs for screening and processing seed. Photo by Greg Houseal.

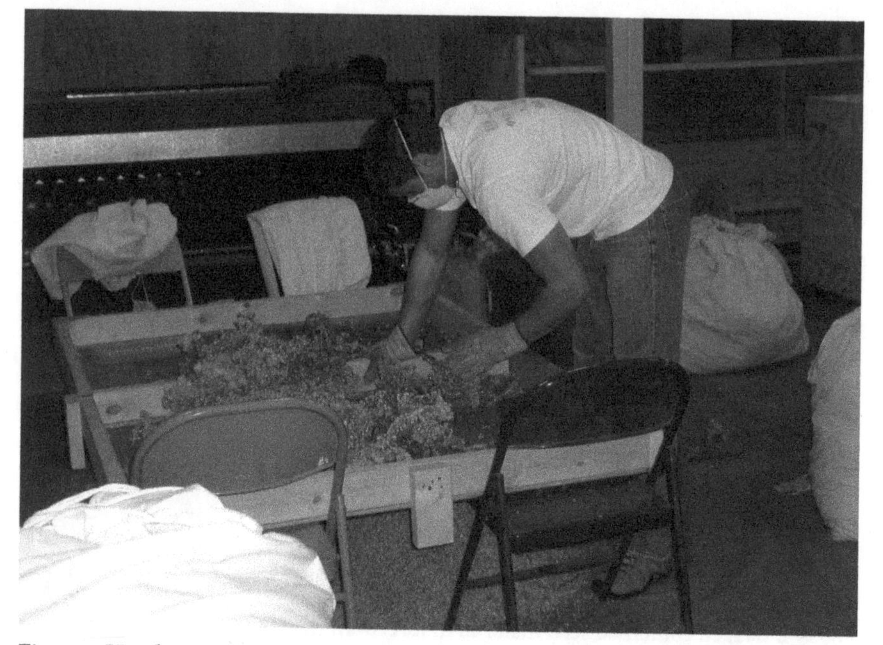

Fig. 15-3. Hand-screening rigid goldenrod to remove seed from hand-clipped seed heads. Photo by Greg Houseal.

seed heads upside down against the inside of a tub, or place them in a bag and shake or beat gently to free the seed. This method has the advantage of minimizing the amount of chaff and inert material in the seed.

The screen-threshing method can be used for hand-clipped and dried seed heads of blazing stars, asters, and goldenrods, and spiderworts can be threshed by rubbing the seed heads over a large screen made of half-inch or quarter-inch hardware cloth, using gloved hands or aluminum scoop shovels. Elevate the screen on sawhorses over a tarp; fluffy seed will float down onto the tarp and can easily be scooped up for further processing (fig. 15-3).

Hand-collected material can be threshed with a variety of machines as well, including hammer mills, huller/scarifiers, and brush machines. Some growers have adapted older combines (with cutter bars and reel removed for safety) for use as stationary threshers of hand-collected material. A debearder (see the next subsection) is effective for threshing the dried seed heads of compass plant and the dried pods of milkweed.

DEAWNING AND DEBEARDING SEED

Many species have appendages, such as long awns or beards on grasses and pappus parachutes on seed of asters, goldenrods, and other composites. Deawning fluffy seed can be accomplished by rubbing it over a small mesh screen with openings just large enough for the seed to pass through, but this is time-consuming and impractical for all but the smallest lots of seed. Standard brass soil sieves are durable and can be used in this way. To begin, make a circular plywood paddle to fit loosely inside the sieves; glue some thin rubber (from an old inner tube, for example) to one side of the paddle, attaching a wooden handle to the other. A small quantity of seed placed on top of the proper screen can be deawned by rotating the rubber-padded paddle back and forth over the seed while applying gentle pressure. Make sure the seed can fall freely through the mesh after deawning, or you risk damaging it. The rubber grips the seed, which improves the deawning action while reducing damage to the seed.

Deawning large quantities of seed requires specialized and sometimes costly equipment. Debearding machines consist of a turning shaft with projecting metal bars housed in a chamber (fig. 15-4). As the chamber fills with seed, the bars work the seed against itself, eventually breaking or rubbing the awns off. It's important, however, to fill the chamber with the proper amount of seed: too little and it's ineffective, too much and the seed can heat up and be damaged. A continuous flow–type debearder works well for larger quantities of seed, but batch-type debearders are also available for smaller quantities. A small, gallon-size Forsberg huller/scarifier is useful for deawning small quantities of seed. This type of machine is very aggressive; only a few seconds of treatment are typically needed (fig. 15-5).

Large seed producers may use a brush machine for removing the pappus from asters, goldenrods, and blazing stars, but this versatile machine actually has many uses. Its basic action is to rub seed and seed heads over a drum screen, or mantle, with rotating brushes. Mantles come in various screen sizes, and a variety of brushes are available. However, the machine can also be used as a deawner or a scarifier and is effective in removing the "cotton" from thimbleweed, threshing seed from hand-collected mints, removing pods from purple prairie clover and leadplant, and deawning smaller quantities of grasses. Heavy canvas beater bars can be installed in place of brushes for a hammer-mill effect.

Cleaning Seed

Cleaning seed involves several techniques to remove plant fragments, dust, and weed seed. Proper cleaning will also remove empty, and therefore nonviable, seed heads. Cleaning techniques involve various ways of sorting, using screens, airflow, and specialized machines. The end goal is high-quality, pure, filled seed, ready for an official seed test and precision planting.

Fig. 15-4. Outflow of big bluestem seed from a Westrup debearder. Photo by Greg Houseal.

Fig. 15-5. A Forsberg huller/scarifier is useful for removing hulls and/or scarifying legume seed; it can also be used to remove parachutes from asters. Photo by Greg Houseal.

SCREENING SEED

Screens are used for sorting by shape and size. Screens are an integral part of fanning mills and air/screen cleaners and are commercially available in a wide range of pore sizes and shapes for these machines. Handheld pan-type screens are handy for small batches. Nested soil sieves are expensive but make excellent durable seed screens. Homemade screens of hardware cloth attached to wood frames are effective for rough cleaning (fig. 15-6). Depending on the application, screens are classified as scalping, grading or sizing, and sifting, as described below.

Scalping removes objects larger, longer, and wider than the desired crop seed. Screens used for scalping have pores larger than the seed. Scalping material through a much larger screen first, then through one closer to seed size, is often more efficient, allowing material to flow more freely through each screen.

Grading sorts the desired crop seed by size. Any given species' seed will contain a range of sizes. Avoid intentionally grading seed intended for restoration plantings, since selection for seed size can happen in one generation (that is, large seeds will give rise to plants with large seeds) and may reduce genetic variability (see chapter 2). Likewise, sometimes seed of a weed species is present that is equal in size and weight to the smaller or larger sizes of crop seed

Fig. 15-6. Rough-screening combined material to remove large stems and leaves in preparation for air/screen cleaning. Photo by Greg Houseal.

present. If this is the case, it may be a necessary trade-off to scalp off a small fraction of the smaller or larger sizes of crop seed in order to effectively remove the weed species.

Sifting is the final screening step. Use a screen with pores just smaller than the seed to allow dust, broken seed, and so on to fall through; the desired seed will remain on the screen.

This series of screening processes is effective in concentrating desired seed and removing most other inert or nonseed material. Additional techniques and equipment, addressed later in this chapter, may be useful in removing same-size seed of other species from crop seed. Likewise, empty, nonviable, but otherwise normal-looking seed will not be removed by simple screening. This "light" seed is removed with airflow, either by winnowing or by aspiration.

WINNOWING SEED

Winnowing uses horizontally moving air to separate heavy from light particles. Winnowing seed in a gentle breeze can remove chaff and light seed very effectively. To achieve more control, place a tarp on the floor and an ordinary box fan at one end of the tarp. Pour seed gently in front of the fan. Heavier seed falls closer to the fan than light seed or empty seed. Fine-tune the process by experimenting with fan speed and distance from the fan. Once you find the most effective combination, continue to pour the seed in front of the fan in a consistent manner. The seed should now be lying somewhat fanned out on the tarp, with the heavier seed nearer to the fan and light or empty seed farther away (fig. 15-7). Using your thumbnail, push down on the seed coats closest to the fan at first, repeating this test as you gradually move away from the fan. Heavy seed will feel firm and resist being crushed with gentle, downward pressure; empty seed, on the other hand, offers little resistance and will crush easily. Determine where the heavy seed ends and the light or empty seed begins, and draw a line through the pile of seed at this point. Clean, heavy seed can then be swept up and stored for planting, while the rest can be discarded.

ASPIRATING SEED

Aspirating uses vertically moving air to suspend particles in a column. Lighter seed is either captured in a pocket of the column, as in a South Dakota seed blower, or blown completely out of the column. Heavier seed drops out of the column. The desired separation is achieved by adjusting airflow in the column.

Fanning mills and air/screen cleaners are designed to combine the screening and aspiration processes and are very efficient once the proper screens and settings have been made (fig. 15-8).

Fig. 15-7. Winnowing Canada anemone seed using a box fan and tarp. Heavy seed falls closer to the fan; lighter and unfilled seed falls farther away. Photo by Greg Houseal.

Fig. 15-8. A Clipper fanning mill can be used to screen and aspirate native seed. Photo by Greg Houseal.

Additional Separation Techniques for Commercial Seed Production

Proper weed control during the growing season—especially the removal of weeds from production fields prior to harvest—is the best way to insure a weed-free finished product. Invariably, weed seed will find its way into the combine and into the crop seed. If weed seed of the same size, weight, and shape as the crop seed is present, it cannot be removed by simple screening and aspirating alone. Additional separating and sorting techniques will be required and may not always be successful. The following machines and techniques are indispensable for the commercial seed cleaning and conditioning necessary to meet federal and state seed laws; however, they require a considerable investment and are not practical for noncommercial use.

LENGTH

Seed can be separated by length in an indent cylinder. An indent cylinder consists of a dimpled drum that rotates nearly horizontally. Seed small enough to fit in a dimple is picked up and dropped by gravity into a trough suspended in the center of the rotating drum. Material too long to fit in the dimples dribbles out the end of the drum. Thus, two fractions are produced: seed and particles shorter than or equal to the dimples and seed and particles longer than the dimples. This is a very effective way to sort foxtail, pigweed, lambsquarters, and so on from longer grass seed. Drums are available with different sizes of dimples.

SHAPE AND SEED-COAT TEXTURE

Seed can be sorted by seed-coat texture with velvet rollers and belt sorters. As seed is dribbled onto an inclined, rotating belt or cylinder, seed with projections, hairs, or rough coats is pulled along on the belt or off over the cylinder, while smooth seed slides off more quickly, creating a separation between smooth and rough seed types. Belt sorters are also effective in separating seed with flat shapes from seed with rounded shapes. Flat seed tends to stay on the inclined belt and is conveyed to a hopper at the end of the rotating belt, while rounded seed tends to roll off the inclined belt and collect in a different hopper. An example of this would be sorting the rounded seed of foxtail from the flatter seed of gray-headed coneflower.

SPECIFIC GRAVITY

Gravity tables are used to sort particles of different sizes and the same density or particles of the same size with different densities. Thus lighter, unfilled, and therefore less dense seed is separated from heavy, filled seed. Gravity tables are most effective on seed that has been graded to a uniform size using grading screens. The gravity table is effective in sorting lambsquarters seed from wild bergamot.

Storing Seed

Proper storage of seed is essential to maintain viability (ability to germinate) and vigor (ability to successfully establish in the field). Three main factors influence the shelf life or longevity of seed during storage: initial seed quality, seed moisture content, and storage temperature combined with relative humidity. Harvest seed at maturity and dry properly, generally to between 5 and 14 percent moisture. In general, a 1 percent increase in seed moisture content can halve seed longevity in storage (Harrington 1972). Once seed has been dried properly, moisture-resistant containers, such as glass or plastic jars, or plastic bags 4 millimeters thick will help protect it from absorbing moisture.

In the short term, seed can be kept in a cool, dry, rodent-proof place for 6 months up to a year before planting. Long-term storage requires controlled conditions of temperature and humidity. A rule of thumb is that the sum of the temperature (in degrees Fahrenheit) and relative humidity should be equal to or less than 100 — for example store seed at 50° F and 50 percent relative humidity. At constant humidity, an increase of 10° F in storage temperature can halve the shelf life of seed (Harrington 1972).

Several other important factors can affect the longevity of stored seed. Inadequate drying may cause mold and mildew during storage. Inert matter left within the seed lot can harbor fungal and insect pathogens, which might damage seed during storage. Cleaning seed properly and thoroughly will extend its viability. Overly aggressive cleaning, however, can damage seed and shorten the longevity of stored seed. Care should be taken with brushing, debearding, and deawning processes not to excessively damage seed. Generally, germination tends to increase slightly in some species stored up to a year after harvest, as dormancy mechanisms break down. Germination then declines over the long term due to seed mortality during storage. Proper storage conditions will slow this decline (fig. 15-9).

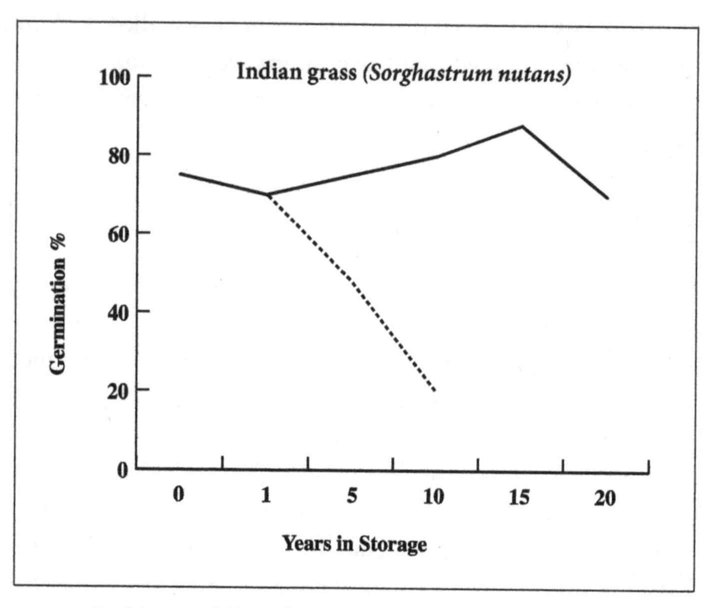

——— Refrigerated Germination %
- - - - - Ambient Germination %

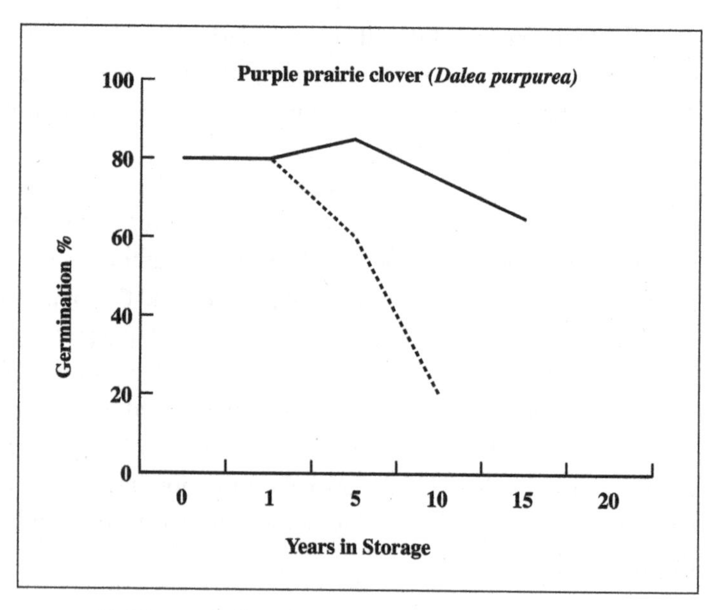

——— Refrigerated Germination %
- - - - - Ambient Germination %

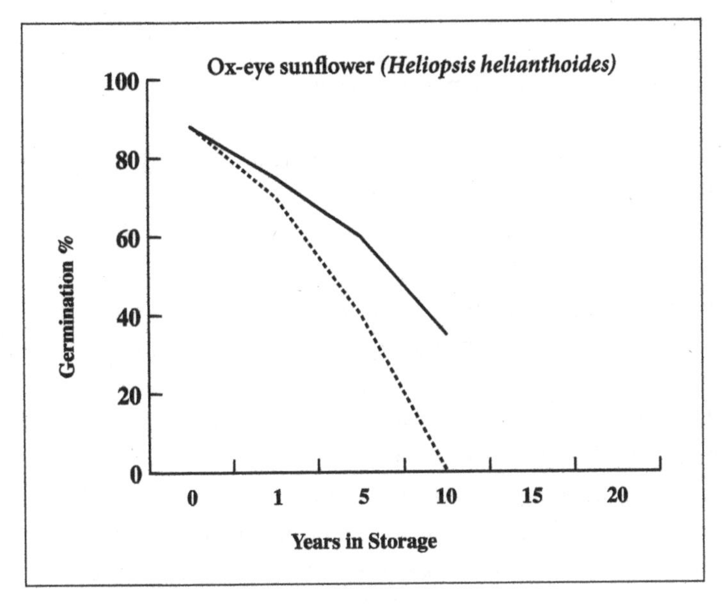

——— Refrigerated Germination %
------ Ambient Germination %

Fig. 15-9. The graphs illustrate the importance of proper storage conditions for preserving seed viability (germination %) over time for Indian grass (facing page, top), purple prairie clover (facing page, bottom), and ox-eye sunflower (above). The solid line represents seed viability in controlled conditions of temperature and humidity (refrigerated). The dashed line represents seed viability at room temperature (ambient) conditions. Note that germination percentage may increase initially as dormancy mechanisms in the seed break down over time. Viability diminishes over time, but much more rapidly when seed is stored improperly. Starched-base grass seed stores well over time. Seed of forb species, such as ox-eye, may not store as well in the long term due to its higher oil content. Forbs with hard seed coats, such as legumes, help preserve viability, as in the purple prairie clover example. Source: Long-term seed storage study, USDA NRCS Manhattan Plant Materials Center, Manhattan, Kansas, courtesy of Rich Wynia.

Summary

- Harvest seed at maturity and dry it properly, generally to between 5 and 14 percent moisture.
- Drying bulk material immediately after harvest is critical for preventing mold and mildew.
- Drying allows immature seed to ripen and aids threshing of seed to maximize seed yield.
- Threshing, debearding and deawning, and brushing are precleaning techniques that remove unnecessary appendages and improve seed flow for further cleaning processes.
- Weed seed of the same size, weight, and shape as crop seed cannot be removed by simple screening and aspirating.
- Simple seed-cleaning techniques requiring minimal investments in equipment are available for rough-cleaning small quantities of seed.
- Air/screen cleaners, debearders and deawners, indent cylinders, velvet rollers, belt sorters, and gravity tables are indispensable for the commercial seed cleaning and conditioning necessary to meet federal and state seed laws; these require a considerable investment.
- Proper storage of seed is essential to maintain viability and vigor.
- Three main factors influence seed longevity during storage: initial seed quality, seed moisture content, and storage temperature combined with relative humidity.
- Properly dried seed can be stored in moisture-resistant containers.
- In general, a 1 percent increase in seed moisture content can halve seed longevity, as can an increase of 10° F in storage temperature.
- A rule of thumb is that the sum of the temperature and relative humidity should not exceed 100.
- Proper storage conditions will slow the natural decline of seed germination that occurs through time due to seed mortality.

16

Propagating and Transplanting Seedlings

GREG HOUSEAL

Overview

Propagating and transplanting seedlings of native species present unique challenges. Most prairie species have seed dormancy mechanisms that must be overcome for good seed germination to occur in a greenhouse environment. Potting containers ideally should accommodate the deep root systems that develop in seedlings. Included in this chapter are tips on greenhouse propagation, potting mediums, containers, light and temperature requirements, and proper watering. The chapter concludes with a discussion of transplanting techniques, including the proper stages of seedling development and the use of a cone-tainer and dibble transplanting system.

Seed Dormancy

Dormancy is an adaptive trait, allowing germination to occur over time and in the proper season. This vital trait prevents the germination of seed at a time that might be suboptimal, or even lethal, for seedling establishment. Staggered germination over time is normal, even with stratification, and should be expected when propagating native prairie species. Successful propagation of native species in a greenhouse requires an understanding of seed dormancy, including how to remove it so seed will germinate. The benefits of removing dormancy are two-fold. First, more seed germinates in a shorter period of time, which means limited and costly greenhouse space is used more efficiently. Second, increased germination means that more individual plants will potentially establish, flower, and reproduce, contributing their genetic diversity to the next generation.

There are two main categories of dormancy: primary and secondary. Primary dormancy occurs when seed is dormant upon dispersal, which is typical of many prairie species. Species with secondary dormancy produce seed with the ability to germinate readily upon dispersal (when it is fresh); however, the seed may enter a dormant state if conditions aren't favorable. Many woodland spring ephemeral species belong to this category of dormancy.

Within the two primary categories, there are several types of seed dormancy, and appropriate strategies are required to remove each type (table 16-1). One type is morphological dormancy. The embryo within the seed is underdeveloped upon dispersal, and warm, moist conditions are generally necessary for maturation (55° to 65° F). Species with this type of dormancy are found in the parsley, buttercup, arum, lily, and iris families, among others (Baskin and Baskin 1998).

Another type, physical dormancy, is due to a physical characteristic of the seed; for example, the seed coat may be hard or waxy or otherwise impermeable to water and gas exchange, thus inhibiting germination. Species in the sumac, legume, geranium, and buckthorn families have these characteristics. Seed with physical dormancy requires some type of scarification to remove these barriers.

Biochemical compounds constitute a third type of dormancy. These compounds may be produced in the seed itself or translocated to the seed from the plant prior to dispersal. Abscisic acid, for example, prevents premature germination of mature seed in the seed head, before dispersal from the parent plant. Concentrations of germination inhibitors, like abscisic acid, decline over time, allowing the seed to germinate. Many seeds may have a combination of dormancy types, which is sometimes called double dormancy. How to remove these developmental, physical, and biochemical inhibitors to germination may not be easily deciphered. If all else fails, gibberellic acid is sometimes effective at inducing germination.

Seed Treatments

SCARIFICATION

Scarification is a technique that simulates the natural disintegration of the seed coat to initiate germination. A hard or waxy coat will not allow the seed to soak up the water needed for germination until the seed coat breaks down. Seed is scarified either through natural processes such as weathering, abrasion, or partial digestion or through artificial techniques. A seed has a natural opening for water uptake, and this weathers or wears away first, especially in a seed with a hard coat, allowing the seed to imbibe water so germination can occur. The trick of scarification, then, is to accelerate the process of weathering this natural opening so the seed can imbibe water, but stopping short of damaging the seed. Some simple scarification techniques are presented here (fig. 16-1).

Table 16-1. Types of Dormancy Associated with Family Groups and Strategies for Breaking Dormancy

Mechanisms of Dormancy	Cause	Removed by
Morphological	underdeveloped embryo	appropriate conditions for embryo development (usually warm, moist conditions)
Examples:	Apiaceae (parsley) Ranunculaceae (buttercup) Araceae (arum) Iridaceae (iris)	
Physical	seed coat impermeable to water/gases	opening of specialized structures (scarification)
Examples:	Anacardiaceae (sumac) Fabaceae (legume) Geraniaceae (geranium) Rhamnaceae (buckthorn)	
Biochemical	germination inhibitors (abscisic acid)	warm and/or cold stratification special light/dark requirements treatment with gibberellic acid
Examples:	many other species	

Source: Adapted from Baskin and Baskin 1998.

Sandpaper blocks can be constructed using rubber cement to glue a sheet of fine-grain sandpaper to each of two flat plywood boards. Lay one sand block on a tray, and use light pressure and a circular motion to move the other sand block on top of a quantity of seed sandwiched between the two blocks. Another variation of this is to line the bottom of a wood box with sandpaper, and make a smaller sand block to use on top of the seed. Adequate scarification is achieved after a minute or two, when the seed begins to look dull.

In percussion scarification, seed is shaken vigorously inside a heavy glass bottle for a few minutes. Allow ample room for all the seed to hit the sides of the bottle. This technique is considered less aggressive and less likely to damage seed than the sand block method. In a variation of percussion scarification, a pneumatic paint shaker was modified by Khadduri and Harrington (2002) to scarify the very hard seed of native locust tree species.

A third technique calls for pouring boiling water over the seed. Use just

Fig. 16-1. Equipment used for seed scarification. Left to right: brass soil sieve with sandpaper-covered paddle, Forsberg huller/scarifier showing paddles and sandpaper-lined canister, and sandpaper-covered boards. Photo by Greg Houseal.

enough to cover them all, then allow them to cool to room temperature, or immerse them in boiling water for 5 to 20 seconds before removing them to rinse and cool. This technique is reserved for certain species and is not broadly recommended. Be sure not to boil the seed! Some species will also require stratification after wet heat. This is effective for New Jersey tea and reportedly for false gromwell.

Commercial scarifiers are also available from seed equipment manufacturers, such as a Forsberg huller/scarifier, which is basically a sandpaper-lined cylinder with metal paddles that turn and agitate the seed. This is a very aggressive method, and only a few seconds are generally needed—precious seed can be reduced to flour if it is left on too long! Seed from commercial producers may already be scarified as part of the cleaning process—if a brush machine was used for dehulling legume seed, for example. Contact your seed vendor or producer to determine whether the seed has been cleaned with a brush machine. See table 16-2 for recommendations on seed scarification.

Table 16-2. Recommended Seed Pretreatments to Induce Germination for Greenhouse Seedling Propagation

	Pretreatment							Sowing			Transplanting
	Scarification	Stratification						Rhizobium Inoculum	Light Required	Planting Depth	Date
Species		Moist	Dry	Warm 68–76° F	Cold 32–45° F	Outdoor Winter Ambient	# Weeks				
Wildflowers											
Canada anemone	x	x		?	x	or x	12			¼"	spring
Thimbleweed	x	x			x		12			¼"	spring
White sage			x		x		12		?	surface	spring
Butterfly milkweed		x			x		4–8			¼"	spring
Sky-blue aster		x			x		8			¼"	spring
Smooth blue aster		x			x		8			¼"	spring
New England aster		x			x		8			¼"	spring
Prairie coreopsis		x			x		12			¼"	spring
Pale purple coneflower		x			x		12		x	cover lightly	spring
Rattlesnake master		x			x		8–12			¼"	spring
Bottle gentian		x			x		12		?	surface	spring
Ox-eye sunflower		x			x		12			¼"	spring
Rough blazing star		x			x		8–12			¼"	spring
Prairie blazing star		x			x		8–12			¼"	spring
Great blue lobelia			x		x		12		x	surface	spring
Wild bergamot			x		x		8–12			surface	spring
Wild quinine		x			x		8–12			¼"	spring

Table 16-2. (*Continued*)

Species	Pretreatment							Sowing			Transplanting
	Scarification	Stratification						Rhizobium Inoculum	Light Required	Planting Depth	Date
		Moist	Dry	Warm 68–76° F	Cold 32–45° F	Outdoor Winter Ambient	# Weeks				
Hairy mountain mint			x		x		12		?	surface	spring
Slender mountain mint			x		x		12		?	surface	spring
Common mountain mint			x		x		12		?	surface	spring
Gray-headed coneflower		x			x		8–12			¼"	spring
Fragrant coneflower		x			x		8–12			¼"	spring
Rosinweed		x			x		8–12			¼–½"	spring
Compass plant		x			x		8–12			¼–½"	spring
Stiff goldenrod		x			x		8–12			¼"	spring
Showy goldenrod		x			x		8–12			¼"	spring
Prairie spiderwort	x	x			x		12			¼"	spring
Ohio spiderwort	x	x			x		12			¼"	spring
Culver's root			x		x		12		x	surface	spring
Golden alexanders	?	x		?	x	[illegible]	[illegible]			¼"	[illegible]

Species	Pretreatment							Sowing			Transplanting
	Scarification	Stratification						Rhizobium Inoculum	Light Required	Planting Depth	Date
		Moist	Dry	Warm 68–76° F	Cold 32–45° F	Outdoor Winter Ambient	# Weeks				
Grasses, warm											
Big bluestem			x		x		—			¼–½"	late spring
Side-oats grama			x		x		—			¼–½"	late spring
Switchgrass		x			x		4			¼"	late spring
Little bluestem			x		x		—			¼"	late spring
Indian grass			x		x		—			¼"	late spring
Prairie cord grass		x			x		4			¼"	late spring
Tall dropseed			x		x		—			¼"	late spring
Prairie dropseed		x			x		4			¼"	late spring
Grasses, cool											
Bluejoint grass			x		x		—			¼"	spring
Woodland reedgrass			x		x		—			cover lightly	spring
Canada wild rye			x		x		—			¼"	spring
Virginia wild rye			x				—			¼"	spring
June grass			x		x		—			cover lightly	spring
Upland wild timothy			x		x		—			cover lightly	spring
Porcupine grass		x			x	or x	12			¼"	spring

Table 16-2. (*Continued*)

	Pretreatment							Sowing			Transplanting
	Scarification	Stratification						Rhizobium Inoculum	Light Required	Planting Depth	Date
Species		Moist	Dry	Warm 68–76° F	Cold 32–45° F	Outdoor Winter Ambient	# Weeks				
Sedges											
Bicknell's sedge		x			x		8		?	surface	spring/fall
Shortbeak sedge		x			x		8		?	surface	spring/fall
Heavy sedge		x			x		8		?	surface	spring/fall
Shrubs											
Leadplant	x	x			x		12	x		¼"	after last frost
New Jersey tea	x	x			x		12			¼"	after last frost
Legumes											
Milk vetch	x	x			x		2	x		¼"	after last frost
White wild indigo	x	x			x		2	x		¼"	after last frost
Cream false indigo	x	x			x		2	x		¼"	after last frost
White prairie clover	x	x	x		x			x		¼"	after last frost
Purple prairie clover	x	x	x		x			x		¼"	after last frost
Showy tick trefoil	x or not	x	x		x			x		¼"	after last frost
Round-headed bush clover	x	x			x		2	x		¼"	after last frost

STRATIFICATION

Stratification is a process whereby seed is placed in a moist medium, such as clean sand or vermiculite, at appropriate temperatures for a period of time. The idea is to mimic the critical conditions necessary for germination that seed is exposed to naturally in the environment after dispersal. Generally, if seed is dispersed in autumn, it may require cold, moist stratification. If dispersed in late spring or early summer, it may require warm, moist conditions or warm followed by cold stratification. Mix seed with an equal amount of moist, sterile sand; sawdust; or vermiculite and place it in a Ziploc bag. Avoid excessive moisture; water should not pool anywhere in the bag. Use vermiculite if working with species adapted to drier conditions to minimize the risk of rot.

Effective temperatures for cold stratification are from 32° to 45° F, with 41° F considered optimum for many species (Baskin and Baskin 1998). Some species require as few as 10 days, others as many as 90 days (table 16-2). A few species, among them American vetch and butterfly milkweed, will germinate at these temperatures, so check the bags weekly to look for emergence of the embryonic root, and plant immediately if this occurs. Some species may germinate best when stratified under natural winter temperature fluctuations (for example, in an unheated building). If sowing seed in flats for outdoor stratification, cover the flats with screen mesh to protect the seeds from being displaced by animals or heavy rains. Cold frames can be used for stratification and to extend the growing season in the spring. Effective temperatures for warm stratification are from 68° to 94° F, with 68° to 76° F optimum for many species with this requirement (Baskin and Baskin 1998).

INOCULATION

Rhizobia are types of nitrogen-fixing bacteria that live symbiotically with the roots of many species, typically forming nodules on rootlets of the plant. They "fix" nitrogen by converting gaseous nitrogen from the air spaces in the soil into plant-available ammonia nitrogen, which directly benefits the host plant. The plant, in turn, provides carbohydrates for the rhizobia. Strains of rhizobia isolated and are commercially available for groups of species, notably in the genera *Amorpha*, *Lespedeza*, and *Dalea*, come as a black powder that is mixed with the seed just prior to planting. Greenhouse-grown seedlings of legumes benefit from rhizobium inoculation. It may be unnecessary to inoculate with rhizobia, however, if seedlings will be planted within a few weeks after germination into native soil where rhizobia naturally occurs.

Mycorrhiza means "fungus roots," which implies a symbiotic relationship between plants and fungi. This is common in many, if not most, plant species. Mycorrhizal fungi occur naturally in healthy soil but may need to be provided for soil that has been fallow, flooded, or eroded over long periods. Sites disturbed by extensive and long-term construction grading or altered by mining will also benefit from inoculation. Commercial inoculum consisting of endomycorrhizal spores, host plant roots, and a sterilized medium is now available, and this can be incorporated into the soil at the time of seeding or transplanting. Inoculation of the site with healthy soil from a different location is another option.

Greenhouse Propagation

Once viable seed is obtained and pretreated to remove dormancy, it's ready to plant in the greenhouse. Critical factors include suitable containers and potting mediums, water, soil temperature, light, and air.

CONTAINERS

Containers should provide good drainage and space for root development, yet be small enough for efficient use of potting medium and bench space. Nurseries interested in retail seedling/plant production will first germinate seed in flats, which take up much less greenhouse space. Seedlings are then transplanted into disposable trays with perforated pull-apart planting cells for retail marketing. Seedlings for transplant into seed-production plots can be placed in a modular reusable "cone"tainer and tray system (fig. 16-2). These cone-tainers are designed to accommodate taproot growth in conifer tree seedlings. They work well for perennial prairie species—particularly those that put down taproots, like compass plant, butterfly milkweed, and *Baptisia* species.

Various sizes of cone-tainers are available in yellow UV-stabilized plastic for longer life. The so-called FIR cells, approximately 6.5 inches deep with a 1-inch diameter at the top, work well for most native species. Each tray accommodates 200 cone-tainers, 100 cone-tainers per square foot, so this is also an efficient use of bench space. Planting dibbles are available that exactly match the size and shape of the various cone-tainers, so transplanting is highly efficient. The cone-tainer and dibble system allows the seedlings to be planted deeply enough that their roots are able to tap into subsoil moisture. Irrigation isn't necessary, especially with spring or fall transplanting when rains are more frequent.

POTTING MEDIUM

A good potting medium should be light enough to allow for good root development, provide adequate drainage, and have enough fertility for seedlings to

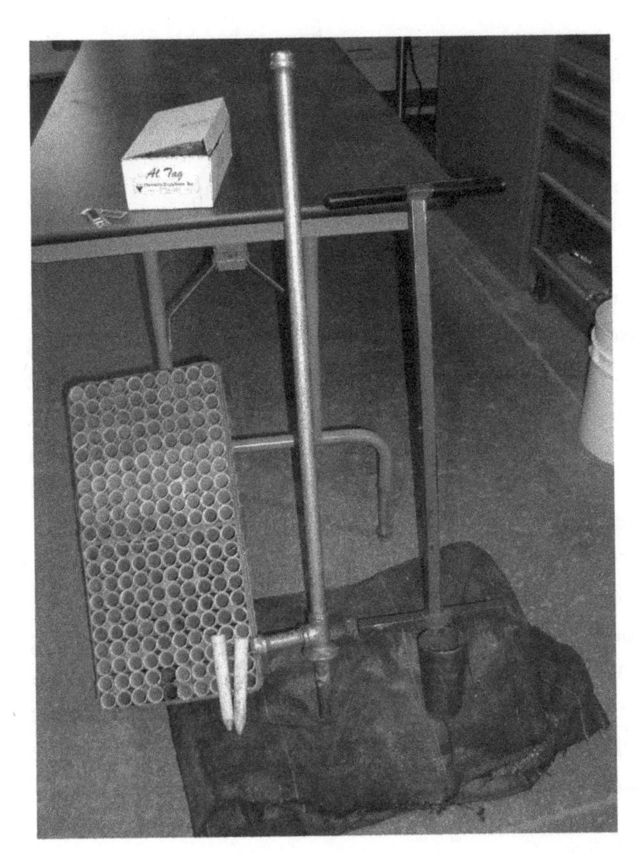

Fig. 16-2. Transplanting equipment. Left to right: aluminum ID tags, Ray Leach cone-tainer rack, planting dibble, bulb cutter, and woven weed barrier. Photo by Greg Houseal.

grow quickly in the greenhouse and become large enough for transplanting in a reasonable amount of time. It should also be sterile, meaning free of weed seed and disease. We recommend a "soil-less" mix (less than 20 percent soil) consisting of 10 percent sterile soil, 10 percent composted (sterile) manure, about 50 percent milled peat moss, and the remainder equal parts perlite and vermiculite. To insure fertility, we use an encapsulated, controlled-release fertilizer (Osmocote, Scotts brand) at the recommended medium rate (8 pounds per cubic yard). Seedlings are susceptible to damage from overfertilization, particularly legumes. We've had good success with the controlled-release fertilizer, and it has the added benefit of continuing to provide fertility after seedlings have been transplanted into production plots. For small-scale home plantings, a premixed and packaged sterile potting soil is suitable.

If mixing on the floor, sweep and vacuum the area prior to mixing to remove seed, debris, and other contaminants. The recipe below makes approximately 1 cubic yard of soil-less mix (enough for 40 trays of 200-cell FIR cone-tainers).

peat moss (4 cubic feet per bag)	2 bags or 8 cubic feet
vermiculite (medium, 4 cubic feet per bag)	½ bag or 2 cubic feet
perlite (4 cubic feet per bag)	½ bag or 2 cubic feet
sterile soil	2 5-gallon buckets
composted (sterile) manure	40-pound bag
Osmocote Plus fertilizer 15-9-12 (180 days)	8 pounds

Screen peat moss, soil, and composted cow manure through a half-inch mesh hardware cloth screen to break up or remove large pieces that will clog and create air pockets as the cone-tainers are being filled. Add the remaining ingredients, mix with shovels on the floor, and fill the trays. A note of caution: All these materials are extremely dusty in their dry form. Be sure to wear a high-quality dust mask when handling and mixing materials. Materials should be moistened with water before mixing to reduce dust.

FILLING AND SEEDING CONE-TAINERS

When filling trays of cone-tainers, tamp the tray on the floor to firm the potting medium and remove large air pockets. Avoid overfilling. Leave about ¾ inch of unfilled space at the top of each cone-tainer. This space acts as a reservoir during watering, allowing the water to seep in slowly, helping to saturate the entire soil column. Water cone-tainers frequently for a day or two before planting to fully hydrate the potting medium. Refill any cone-tainers that may have settled excessively.

Attempt to sow several seeds per cone-tainer. If the seed has been mixed with damp sand or another medium for stratification, it may be impossible to distinguish small seed from the medium. If this is the case, place the damp seed/medium mixture in a shallow dish, and mix thoroughly to distribute the seed evenly within the medium. Use a small, flat implement (the point of a knife or a wooden popsicle stick) to place a small amount of mixture in each cone-tainer. Getting an appropriate number of seeds per cell is guesswork with tiny seeds, but experience will improve efficiency. Thinning may be necessary if too many seeds germinate in each cell. Blank cells will result if no viable seed was planted. For larger seeds with high purity and viability (90 percent or more), one to three seeds per container is adequate. Increase accordingly for seed of poor or unknown quality. Cover with no more than ⅛ to ¼ inch of soil for most species. Very tiny seeds should not be covered. Additional information and precautions on sowing seed of specific species can be found in the following subsections on light, temperature, and watering.

LIGHT

Natural light is best for growing native prairie species and should be sufficient for seedling establishment in the greenhouse from mid March through mid September. Sow seed in early February, and expect germination and emergence to occur over a 2- to 6-week period. Greenhouse-grown seedlings grow well with only natural light through March and April and into May, when transplanting into production plots begins. Keep in mind that some species require light for germination. These are typically small-seeded species, including but not limited to Culver's root, mountain mints, grass-leaved goldenrod, Joe Pye weed, great blue lobelia, and white sage. These do best if sprinkled on top of the soil surface and kept continually moist until the seed leaves (cotyledons) are evident.

TEMPERATURE

Germination will occur throughout a range of temperatures but will be slower when temperatures are less than optimal. The risk of fungal pathogens and rot increases if seed does not germinate and is nondormant. Warm-season grasses and legumes germinate best in soil warmer than 70° F. Cool-season grasses and many forb species will germinate more readily in cool soil temperatures of 40° to 50° F and may cease germination at temperatures above 77° F. Soil temperature in the cone-tainers fluctuates with greenhouse air temperature (72° F daytime, 60° F nighttime).

Pulses of emergence occur on sunny days with some species, presumably because an optimum soil temperature has been reached from solar heating. Covering cone-tainers with translucent plastic will increase soil temperature and improve germination of species that require warm soil. Use this technique with caution. Lethal temperatures can occur quickly under the plastic with full summer sun. Plastic should be removed as soon as the first seedlings emerge to avoid overheating new seedlings. Cooler soil temperatures can be achieved by setting trays of cone-tainers on the floor. If you are sowing seed in flats, precise regulation of soil temperature can be achieved with propagation mats. Propagation mats are placed under the flats, then plugged into an electrical temperature control box. Soil temperature in the flats is regulated by a soil temperature probe from the control box inserted into the potting soil of one of the flats. These are commercially available at a reasonable cost from nursery or greenhouse supply companies.

WATERING

Proper watering is critical to greenhouse propagation. It's important to keep the soil surface moist until germination has occurred, especially for small seed that requires light and is sowed directly on the soil surface. An automated mist system is helpful during this stage of propagation. If using a wand or a watering can, use a sprinkler head that produces small, gentle droplets and low pressure so that watering doesn't dislodge seed, forcing it deeper into the soil or splattering it out of the containers (fig. 16-3). The large proportion of dried and milled peat in soil-less mixes sometimes impedes initial wetting of the potting medium, and water will not be readily absorbed below the surface of the medium. This can be remedied by thoroughly wetting the potting medium during the mixing process. A small amount of dish soap added to the water will act as a surfactant and aid wetting when watering.

Once plants are established, they should be watered thoroughly at least once a day, insuring that the entire soil column is moistened, but allowing the soil to drain and the surface soil to begin to dry slightly. More frequent watering (2 to 3 times daily) is required on hot and sunny days and for larger plants with fibrous roots that fully occupy the container. Underwatering will allow the lower portion of the soil to dry out, and root growth will be stalled. Containers that are overfilled with soil are prone to underwatering, since the water can't

Fig. 16-3. When using a wand or a watering can, use a sprinkler head that produces small, gentle droplets and low pressure so that watering doesn't dislodge seed. Photo by Greg Houseal.

pool on the surface and gradually soak in. Likewise, overwatering saturates the soil, depriving the root zone of air and creating conditions conducive to decay. A potting medium and containers that allow for proper drainage will help prevent overwatering.

Excessive watering can also damage shoots. Healthy-looking plants will suddenly fall over, appearing to be cut off at the soil level. Known as damping off, this is caused by a fungus. Legumes are particularly susceptible to this condition, but it can affect other species as well, especially if they are planted too densely. Sprinkling a layer of perlite over the top of the soil surface after seeding will help dry the soil surface and wick water away from the stems. Maintaining good air circulation will evaporate excess water from stems and soil surface and minimize the risk of damping off. Thinning may be necessary to improve air circulation. Watering containers from below by setting them in a pan of water until soil wicks up moisture will also help. Washing and sterilizing containers, benches, and equipment and using a sterile potting medium will also help reduce the risk of damping off.

Transplanting Seedlings

SEEDLING DEVELOPMENT AND TIMING

The key to successful transplanting of native perennial species is strong root development. Ideally, roots should fully occupy the entire soil column, so that when the plant is removed, the soil and roots remain intact as a plug (that is, they retain the shape of the cone-tainer). Grasses and forbs with fibrous roots form beautiful plugs after a few weeks of growth and present little challenge in transplanting. Species with taproots (*Baptisia* species, compass plant, *Desmodium* species, butterfly milkweed) develop a thickened, fleshy taproot within a few weeks after germination in a greenhouse, but the taproot itself may not be enough to hold the plug intact when transplanting. Allow growth to continue until the taproot reaches the bottom of the cone-tainer. The taproot will air-prune (it can't grow further), and growth of fine lateral roots will be stimulated. These fine lateral roots will help considerably in holding the plug intact when the seedling is removed from the cone-tainer for transplanting. Slower-growing forbs and shrubs require more time for roots to develop adequately for transplant (figs. 16-4 and 16-5).

Seedlings are prone to top-kill when transplanted directly from the greenhouse into the field. Greenhouse-grown seedlings are pampered; they have been protected from drying wind, harsh sun, and herbivories. Robust, greenhouse-grown seedlings of many native species can tolerate the stress and will regrow

Fig. 16-4. Crew transplanting plugs into weed barrier using the cone-tainer and dibble system. Photo by Greg Houseal.

quickly if they have strong root development and adequate soil moisture. A better approach, however, is to acclimate seedlings gradually to outdoor conditions with a process called hardening off. A week before transplanting, set the flats or trays outside for a few hours each day, from mid morning to mid afternoon, in a place sheltered from strong winds and full sun. The idea is to acclimate the plants gradually to outdoor conditions of wind and sun. Strong winds and heavy rains should be avoided. Another option is to move flats or trays into a cold frame (an unheated greenhouse), and roll up the sides or open the side vents to allow natural airflow and some direct sun to the plants.

The ideal time for transplanting is in the spring, after the last frost-free date for your region. Rains are more reliable at this time of year, the sun is less intense, and plants have the entire growing season to establish and flourish. If transplanting during the summer months, check to insure adequate soil moisture. Be prepared to water regularly and deeply until plants are established. Transplanting in the fall (early to mid September) may be another option, if root development is strong enough to allow the seedlings to survive the winter months (table 16-3).

Fig. 16-5. Native seedlings removed from cone-tainers show well-developed root plugs ready for transplanting. Photo by Greg Houseal.

HOW TO TRANSPLANT SEEDLINGS

Soil should be firm in all cases. If you are dibbling into bare soil, the soil should be rolled or packed to prevent dibbled holes from collapsing. Likewise, very dry soil resists dibbling. It may be necessary to sprinkle the area the day before, or wait a day or two after a soaking rain. Just prior to transplanting, liberally water seedlings in cone-tainers to fully saturate the root plugs. This will make it much easier to remove the plugs from the cone-tainers and will provide extra moisture to the root zone after transplanting. Hold the cone-tainer upside down firmly in one hand, and rap the rim sharply with the palm of the other hand, using a flick of the wrist. The plug should slide out easily; repeat if necessary. (If plugs do not hold their shape upon removal, either the roots are not adequately developed, or too much force is being used. Transplant success drops significantly if this happens!) Slide the intact plug into the dibbled hole; it should just fit, with the top of the root plug just at or slightly below the soil level. Pinch the soil firmly around the top of the plug to seal in moisture, taking care not to bury the base of the shoot. The lighter soil-less mix can actually wick moisture away from the roots if they are left exposed. Be sure the dibbled holes are deep

Table 16-3. Propagation and Transplanting Recommendations for Individual Species

Species	Recommended Establishment Method	Seeding Time Direct Seed	Transplanting Time		Seeding Depth
			Seedlings	Division	
Wildflowers					
Canada anemone	seed, DIV	dormant	spring	spring/fall	¼"
Thimbleweed	DS	—	spring		¼"
White sage	seed, DIV	—	spring	spring/fall	surface
Butterfly milkweed	seed	dormant	spring	fall	¼"
Sky-blue aster	seed	—	spring		¼"
Smooth blue aster	seed	—	spring		¼"
New England aster	seed	dormant	spring	spring/fall	¼"
Prairie coreopsis	seed	dormant	spring	spring/fall	¼"
Pale purple coneflower	seed	dormant	spring		cover lightly
Rattlesnake master	seed	dormant	spring	spring	¼"
Bottle gentian	seed	—	spring		surface
Ox-eye sunflower	seed	dormant	spring		¼"
Rough blazing star	seed, DIV	—	spring	fall	¼"
Prairie blazing star	seed, DIV	—	spring	fall	¼"
Great blue lobelia	seed	—	spring		surface
Wild bergamot	seed	dormant	spring	spring/fall	surface
Wild quinine	seed	dormant	spring		¼"
Hairy mountain mint	seed	dormant	spring		surface
Narrow-leaved mountain mint	seed	[illegible]	spring		surface

Species	Recommended Establishment Method	Seeding Time Direct Seed	Transplanting Time		Seeding Depth
			Seedlings	Division	
Common mountain mint	seed	—	spring		surface
Gray-headed coneflower	seed	dormant	spring		¼"
Fragrant coneflower	seed	dormant	spring		¼"
Rosinweed	seed	—	spring		¼–½"
Compass plant	seed	—	spring		¼–½"
Stiff goldenrod	seed	dormant	spring	spring/fall	¼"
Showy goldenrod	seed	—	spring		¼"
Prairie spiderwort	seed, DIV	—	spring	spring/fall	¼"
Ohio spiderwort	seed, DIV	—	spring	spring/fall	¼"
Culver's root	seed, DIV	—	spring	spring/fall	surface
Golden alexanders	DS	dormant	spring		¼"
Grasses, warm					
Big bluestem	seed	late spring	late spring		¼–½"
Side-oats grama	seed	late spring	late spring		¼–½"
Switchgrass	seed	dormant	late spring		¼"
Little bluestem	seed	late spring	late spring		¼"
Indian grass	seed	late spring	late spring		¼"
Prairie cord grass	seed, DIV	late spring	late spring	spring	¼"
Tall dropseed	seed	late spring	late spring		¼"
Prairie dropseed	seed, DIV	—	late spring		¼"

Table 16-3. (*Continued*)

Species	Recommended Establishment Method	Seeding Time	Transplanting Time		Seeding Depth
		Direct Seed	Seedlings	Division	
Grasses, cool					
Bluejoint grass	seed, DIV	fall/early spring	spring	spring/fall	¼"
Canada wild rye	seed	fall	spring		¼"
Virginia wild rye	seed	fall	spring		¼"
June grass	seed	—	spring		cover lightly
Legumes					
Milk vetch	seed	dormant	after last frost		¼"
White wild indigo	seed	dormant	after last frost		¼"
Cream false indigo	seed	dormant	after last frost		¼"
White prairie clover	seed	dormant	after last frost		¼"
Purple prairie clover	seed	dormant	after last frost		¼"
Showy tick trefoil	seed	dormant	after last frost		¼"
Round-headed bush clover	seed	dormant	after last frost		¼"

Note: Seed = sow seed in containers in greenhouse, DIV = division by roots/corms/rhizomes, DS = direct seed.

enough to comfortably receive the full depth of the plug. Plugs forced into a hole that is too small or too shallow will often pop out of the ground after a good rain, exposing the root collar.

Summary

- Dormancy is an adaptive trait, allowing germination over time and in the proper season.
- Successful propagation of native species in a greenhouse requires an understanding of seed dormancy, including how to remove it so seed will germinate.
- Scarification is a technique that accelerates the natural weathering some seeds require to remove barriers to allow for water absorption and subsequent germination.
- Stratification is a process whereby seed is placed in a moist medium at appropriate temperatures for a period of time to mimic the critical conditions necessary for germination.
- Effective temperatures for cold stratification are from 32° to 45° F, with 41° F considered optimum for many species.
- Effective temperatures for warm stratification are from 68° to 94° F, with 68° to 76° F optimum for many species with this requirement.
- Rhizobial bacteria and/or mycorrhizal fungi occur naturally in healthy soil and may need to be provided in some situations.
- Containers should provide good drainage and space for root development, yet be small enough for efficient use of potting medium and bench space.
- A good potting medium should be light enough to allow for good root development, provide adequate drainage, and have enough fertility for seedlings to grow quickly.
- Natural light is best and sufficient for seedling establishment in the greenhouse from mid March through mid September.
- Warm-season grasses and legumes germinate best in soil warmer than 70° F; however, cool-season grasses and many forb species will germinate more readily in soil temperatures of 40° to 50° F.
- It's important to keep the soil surface moist until germination has occurred, especially for small seed that requires light and is sowed directly on the soil surface.
- Once plants are established, they should be watered thoroughly at least once a day, insuring that the entire soil column is moistened, but allowing the soil to drain and the surface soil to begin to dry slightly.

- Excessive watering may create conditions for damping off, whereby otherwise healthy shoots fall over, appearing to be cut off at the soil level.
- Acclimate or harden off seedlings gradually to outdoor conditions of wind and sun a week before transplanting.
- The key to successful transplanting of native perennial species is strong root development.
- The ideal time for transplanting is in the spring after the last frost-free date for your region.

Epilogue:
The Future of Tallgrass Restoration

DARYL SMITH

Writing this guide was a marvelous experience, both invigorating and reflective. As the four of us wrote various sections and critiqued one another's writings, we engaged in some very stimulating discussions. In addition, it caused me to pause and reflect on my nearly 40 years of experience with prairie restoration and reconstruction.

I grew up prairie-deprived and never knowingly encountered a prairie until I was 26 years old. I first viewed prairie as a botany graduate student at the University of South Dakota on a field trip with Ted Van Brugen. I think it was love at first sight; the prairie flowers were beautiful, and the images of original prairie landscape that I created in my mind were awesome. I became a frequent visitor to prairies of the Vermillion area that summer. In later years, as I reflected on that initial introduction to prairie, I recalled being frustrated about missing prairies for more than two decades of my life.

My prairie passion became dormant as I pursued a doctorate at the University of Iowa, investigated "the role of pectin methylesterase in abscission," and joined the faculty at the University of Northern Iowa. The prairie flame began to flicker as Paul Shepard, keynote speaker for the first Earth Day at UNI, described prairie restoration work at Knox College. Soon thereafter, Glen Crum, a nontraditional graduate student, persuaded me to visit a sand prairie northwest of Cedar Falls. The visit was fantastic; the flame of my passion for prairie rekindled and burned brightly.

I began to delve into prairie reconstruction in 1972, when the UNI Biological Preserves Committee decided to create microcosms of Iowa plant communities on campus. I jumped at the opportunity to take the lead in reconstructing a tallgrass prairie. I was convinced that a prairie on campus would increase opportunities for students to experience this endangered ecosystem.

With the aid of a 1972 summer fellowship, I thoroughly prepared for the campus prairie reconstruction. I sought the advice of Paul Christiansen and Roger "Jake" Landers, read about the reconstructions by Peter Schramm at Knox College and Ray Schulenberg at the Morton Arboretum, visited local prairie remnants and Curtis Prairie at the University of Wisconsin Arboretum.

The Third Midwest (later North American) Prairie Conference at Kansas State University in the fall of 1972 was great. It was a heady experience to listen in as Ray Schulenberg, Paul Christiansen, Peter Schramm, and others discussed prairie reconstruction. They were breaking new ground, and I felt as if I was on the cutting edge of a new world.

I was anxious to begin my first prairie reconstruction. In my haste, I made mistakes. Some could be corrected, but others persisted to remind me that patience is one of the virtues of reconstruction. However, the project has been rewarding. Early on, UNI became known as the university with a prairie on campus, introducing students of all ages to prairie. Over the next decade, I increased my prairie advocacy—I helped preserve prairies, visited and studied prairies, honed my skills in prairie reconstruction, used fire management on prairies, developed prairie studies courses, and promoted prairie statewide. My CB handle was the "prairie prophet." I directed the Twelfth North American Prairie Conference in 1990 and a decade of biennial Iowa prairie conferences.

The experience and expertise I had gained in prairie reconstruction resulted in UNI being asked to develop a statewide program to assist Iowa counties in implementing and maintaining programs in Integrated Roadside Vegetation Management (IRVM). The cornerstone of IRVM is the use of native prairie species to out-compete weeds, thus reducing mowing and herbicide use. We initiated and developed a program to produce increased quantities of source-identified prairie seed at an economically competitive price. The success of Iowa's IRVM programs and the seed-production model vaulted the state into national prominence in roadside vegetation management. I felt like I had started rolling a snowball downhill, and I was now trying to hang on.

The national exposure enabled us to secure funding through the Federal Highway Administration to develop a center for native roadside vegetation management. Thirty years after I borrowed a Nesbitt drill to seed the campus prairie, we moved into our own facility with a program focused on roadside management and prairie reconstruction. Since that time, external funding has permitted us to add staff and to expand our activities at the Tallgrass Prairie Center. All are committed to providing prairie experiences for as many people as possible, and all are convinced that we are restoring a national treasure.

Undoubtedly, prairie restoration and reconstruction will increase in the next 40 years. Ecosystem alteration in the next several decades is inevitable as the human population and technological capability increase. I feel it is essential that prairie ecosystems be restored or replicated to retain self-sustaining ecosystems and insure that people continue to know and appreciate prairie. Interest in prairie remains high with prairie plantings and reconstructions by landowners; conservation organizations; local, state, and federal agencies; roadside managers; and prairie enthusiasts. In addition, more nurseries and landscaping companies are specializing in planting prairies and/or selling prairie seed and plants.

The Tallgrass Prairie Center interacts with all these entities involved in restoration. We can work with practitioners and restoration ecologists to exchange information and test techniques to generate principles and guidelines that advance prairie restoration. And equally important, we can help advance prairie restoration by applauding their dedication and efforts while sharing the energy and passion that they generate.

Glossary

Annual: A plant that germinates, matures, flowers, sets seed, and dies in a single year.

Aspect: The direction that a slope faces.

Aspirating: Using vertically moving air to suspend particles in a column; lighter seeds are either captured in a pocket of the column or blown completely out of the column.

Assessment: A measure of the quality of a remnant.

Auricle: Appendage extending from the leaf collar.

Awn: A hairlike or bristlelike appendage on a plant or seed.

Beard: Fine hairs on a seed.

Biennial: A plant that germinates and grows in the first year and resumes growth, matures, flowers, sets seed, and dies in the second year.

Biodiversity: All life forms within an area, habitat, ecosystem, biome, or the entire earth.

Biomass: The total number of organisms in a given area measured as live, dead, dry weight, or energy (calories).

Biotic community: An assemblage of living organisms that occur together.

Blacksoil prairie: A prairie that typically has a dark topsoil up to 2 feet deep that is rich in organic matter. Beneath this, a clay subsoil retains moisture and is rich in minerals. The lowest layer of subsoil is the mineral-rich glacial till that was left behind by the last glaciers. It is typically a mixture of finely pulverized rocks, clay or sand, and loess (fine, wind-borne debris).

Bog: A poorly drained, usually acid area rich in accumulated plant material, frequently surrounding a body of open water and having a characteristic flora (such as sedges, heaths, and sphagnum).

Broadcast seeder: A seeding device that sows the seed over the soil surface.

Broadcast seeding: A seeding method that involves random dispersal of seed over an area by hand or by mechanical means.

Bulk-harvest: To collect a mixture of seed, chaff, leaves, and stems of species present in a prairie.

Cambium: A layer below the inner bark that is capable of active cell division, producing xylem to the inside of the plant and phloem to the outside.

Canopy: The uppermost layer of vegetation in a plant community that creates a shade-forming layer.

Caryopsis: A small, dry, one-seeded fruit in which the ovary wall remains joined with the seed in a single grain, as in barley, wheat, corn, and rice.

Ciliate: Having very small, fine hairs.

Climax community: An assemblage of plants and animals which, through the process of ecological succession—the development of vegetation in an area over time—has reached a steady state.

Clonal growth: An organism derived from asexual or vegetative multiplication that originated from a single parent.

Clonal spread: A group of individuals growing from rhizomes or stolons outward from a single parent plant.

Clopyralid: A synthetic growth hormone herbicide that kills broadleaf plants by causing uncontrolled and disorganized growth, typically used to control thistle and clover species.

Coefficient of conservatism: A numerical score assigned to each plant species in a local flora that reflects the likelihood that a species is found in natural habitats.

Compacted soil: Compressed soil that inhibits the penetration of plant roots.

Cone-tainer: A cone-shaped receptacle used to germinate individual seeds and allow for growth prior to transplanting.

Conservative species: Species that serve, almost exclusively, to distinguish intact natural areas from disturbed landscapes.

Control method: Method used to reduce a population size or eliminate a group of organisms by various techniques such as chemical application, mechanical operations, or hand-weeding.

Cool-season grass: A grass that grows during the spring, then flowers and sets seed in late spring or early summer.

Cotyledon: The first seed leaf or leaf pair to appear after germination.

Cover crop: A quick-growing crop, such as an annual cereal crop, seeded with perennial species to reduce soil erosion and weed growth.

Crop residue: Plant material left in the field after harvest that includes leaves, stalks, and stubble (stems).

Cultipacker: A heavy implement pulled by a tractor that rolls across the soil, firming the seedbed and packing the seed.

Cultipacking: Rolling the soil surface to reduce the size of dirt clods and to remove large air pockets to create a smooth, firm seedbed. Rolling the surface after sowing can improve seed placement in the soil and seed-to-soil contact.

Cultivars: Varieties of grass and forb species that have been selectively bred for desired characteristics, abbreviation for "cultivated variety."

Culturally modified prairie: A prairie changed by human activity such as grazing, mowing, or haying.

Deawning: Removing the awn from a seed.

Debearding: Removing the beard from a seed.

Degradation: The process of breaking down into smaller, simpler parts, for example, when an herbicide compound applied to vegetation breaks down into its element components in the soil.

Degraded prairie remnant: A remnant that is heavily disturbed, modified, or damaged.

Dibble: A small hand implement used to make holes in the ground for plants, seeds, or bulbs.

Diversity: The number and frequency of species in one location.

Dormant seeding: Seeding during a time of the year when plants are not actively growing, such as late fall and winter.

Drill seeding: Sowing seed with a mechanized implement that plants seed in the soil and regulates both planting depth and application rate.

Ecosystem: A system that includes all living organisms (biotic factors) in an area as well as its physical environment (abiotic factors) functioning together as a unit.

Edge effect: Plant/animal population responses in areas where two or more habitat types come together.

Edge-to-area ratio: A ratio determined by dividing the area of a reserve by its perimeter (edge).

Emergent plant: A plant arising through the soil surface from a germinated seed.

Establish: The ability of a seed to germinate, emerge, and take root at a given location.

Extirpated species: A species that has been eliminated from an area.

Fen: An alkaline wetland characterized by water percolating from under the ground.

Forb: An herbaceous flowering plant.

Friable: Easily crumbled, crumbly, such as crumbly soil particles that are ideal for root growth of plants.

Frost seeding: Sowing seed on top of frozen ground in late winter or early spring, when repeated freezing and thawing of the soil will cause surface cracks and promote the incorporation of seed into the soil.

Genetic diversity: The variability of genetic characteristics of a species.

Genotype: The genetic constitution of a cell, an organism, or an individual, usually with reference to a specific characteristic under consideration.

Germinate: To initiate the growth of a seed.

Germination: The initiation of growth and development of a seed.

Girdling: Removing a shallow ring of bark around a tree, thus severing phloem tissue so nutrients cannot be transported to the roots, resulting in the death of the tree.

Glyphosate: An herbicide absorbed through leaves or a cut stump that inhibits an enzyme involved in the synthesis of amino acids and results in plant death.

Granivores: Animals and insects that feed on the seeds of plants.

Granivory: Seed predation.

Grassland: An area dominated by plants in the Poaceae or grass family.

Hard dough: At this stage, the seed kernel is well developed, and firm thumbnail pressure will dent it.

Herbaceous: Nonwoody, used to describe plants.

Herbicide: A chemical compound used to kill plants.

Hydric soils: Somewhat poorly drained, poorly drained, and very poorly drained soils that typically have standing water for part or most of the growing season.

Hydrologic gradient: Variation in soil moisture along two regimes.

Hydrology: The study of the distribution, movement, and quality of water.

Hydroseeding: Spraying a slurry mixture of seed, water, and mulch directly onto the ground.

Interseeding: Sowing seed into a stand of established vegetation.

Invasive species: Any species, including its seeds, eggs, spores, or other biological material capable of propagating that species, that is not native to a given ecosystem and whose introduction is likely to cause economic or environmental harm or harm to human health.

Ligule: A membranous appendage arising from the inner surface of a leaf at the junction with the leaf sheath.

Local ecotype seed: Seed presumed to contain genetic adaptations derived from local remnants.

Mature prairie reconstruction: A stage in the development of planted prairie where most of the native plants are well established and reproducing.

Medium dough: At this stage, moderate pressure with a thumbnail will dent the seed.

Mesic soil: Well-drained and moderately well drained loamy soil.

Milk stage: At this stage, the tissue within the seed is milky and not firm.

Morphological seed dormancy: A condition in which seeds have fully differentiated embryos that are not yet fully developed and need to grow more before germination.

Mosaic seeding: Seeding different species or mixtures of species onto specific areas of the planting site where they are most likely to establish and persist.

Mycorrhiza: A symbiotic association between a fungus and the roots of a plant.

Native species: A species whose presence in a given region or ecosystem is the result of only natural phenomena, with no human intervention.

Nonnative species: A species living outside of its native distribution.

No-till drill: A seeding device that can sow seed into sod without tilling the soil.

Nurse crop: A quick-growing annual species that is included in the seed mix with perennials to stabilize the soil and reduce weeds.

Opportunistic species: Species characterized by high reproduction rates, rapid development, early reproduction, small body size, and uncertain adult survival.

Pappus: A cluster of bristles attached to the upper part of the ovary (a part of the female reproductive organ of a flower) that may aid in seed dispersal.

Pathogenic fungi: Microscopic fungi producing or capable of producing disease.

Perennial: A plant that lives for 3 or more years.

Perlite: An amorphous volcanic glass that is a component of soilless growing mixes, where it provides aeration and optimum moisture retention.

Pesticide: A chemical compound that prevents, destroys, or repels pests.

Phenology: The study of recurring changes in an organism's growth and their relation to climate and changes in the season.
Phloem: The nutrient-conducting tissue of vascular plants located below the inner bark.
Photo point: A permanent location for photographing the same landscape perspective over time.
Physical seed dormancy: A condition in which the seed coat prevents the seed from absorbing water, thus inhibiting germination.
Plant count: Identifying and counting plants within a designated sample area.
Plant density: The number of plants per unit area, often expressed as an average of plants per square meter or per square foot.
Plant frequency: The proportion of samples of the total samples that a species or group of species is detected in.
Prairie: A treeless plant community dominated by grasses and forbs native to the region.
Prairie reconstruction: The act of restoring sites where prairie species no longer exist.
Prairie remnant: A remaining portion of a plant community dominated by native grasses and forbs on an undisturbed soil profile.
Prairie restoration: An umbrella term used to describe both reconstruction and restoration efforts.
Predation: The capture and consumption of an organism (prey) by another organism (predator).
Prescribed fire: Controlled application of fire to wildland fuels in either their natural or modified state, under specified environmental conditions that allow the fire to be confined to a predetermined area and at the same time to produce the intensity of heat and spread required to attain management objectives.
Primary seed dormancy: A condition in which seed germination is delayed by physical, mechanical, chemical, morphological, or physiological factors of the seed.
Propagation: Multiplication of plants by seed or vegetative means.
Propagule: Any part of an organism, produced sexually or asexually, that gives rise to a new organism, such as a stem cutting, leaf, stolon, rhizome, or seed of a plant.
Pure live seed (PLS): The percentage of purity of a seed lot multiplied by the percentage of viability as determined by a seed test and expressed as a weight (pounds).
Purity: A measure of pure, unbroken crop seed units as a percentage by weight of the seed lot.
Quadrat: A frame of known area that the investigator places on the ground in order to sample the vegetation inside the frame.
Randomized sampling: A technique used to select sample sites by insuring that all locations of the planting and all individuals of the population have an equal chance of being sampled.
Reference point: A permanent object in a photograph giving orientation and scale to images over time.

Reference remnant: A prairie remnant that is used as a model reference for species selection and seed source to restore or reconstruct a prairie within the same region.

Refugia: An area of equivalent habitat undisturbed by natural or human-induced change that provides a safe haven for flora and fauna.

Regional source seed: Seed pooled from several remnant populations.

Remnant source seed: Seed that is collected directly from (and shares genetic identity with) a prairie remnant or seed produced from seed originally collected from a prairie remnant.

Rhizobium bacteria: Soil bacteria that fix nitrogen after becoming established inside the root nodules of legumes. The rhizobia cannot independently fix nitrogen and thus require a plant host.

Rhizome: A creeping stem growing horizontally under the soil surface that often sends out roots and shoots from its nodes.

Rosette: A dense, radiating cluster of leaves at or near ground level.

Sample bias: A glitch in the sampling procedure producing results that do not represent the actual condition in the field.

Sampling: The random selection of discrete observations (samples) intended to yield knowledge about a community or population of concern.

Savanna: A tropical, subtropical, or temperate grassland containing scattered trees and drought-resistant undergrowth.

Scarification: A technique that accelerates the natural weathering some seeds require to remove barriers to allow for water absorption and subsequent germination, usually accomplished by scratching the seed coat to allow water to penetrate.

Secondary seed dormancy: A condition where temperature extremes, prolonged light or darkness, water, or oxygen stress will prevent seed germination after the seed has imbibed water.

Secondary succession: Ecological succession that occurs in habitats where the previous community has been destroyed or severely disturbed, such as following forest fire, abandonment of agricultural fields, or epidemic disease or pest attack.

Sedges: Grasslike plants in the genus *Carex* that have pointed leaves, triangular solid stems, and minute flowers in the spikelets.

Sedge meadows: Plant communities dominated by species in the genus *Carex* that grow in saturated soil.

Sedimentation: The process of deposition of sediments—material that has been picked up and transported by wind, water, or ice—in a variety of environments.

Seed dormancy: A condition of plant seeds that prevents germination until optimal environmental conditions for their germination exist.

Seeding rate: The number or weight of seeds sowed per unit area, measured as seeds per square foot or pounds per acre.

Seed mortality: Seed death.

Seed source: The original location (genetic identity) from which the seed was collected or derived.

Seep: A wetland where water slowly flows from the ground.

Slope: A piece of ground that is not flat and is measured in ratios, such as 3:1, indicating 3 feet of run for each foot of rise.

Soft dough: At this stage, a seed kernel is well formed and filled with starch. When squeezed, there is no milky fluid, only a rubbery, doughlike substance. This stage lasts about a week to 10 days.

Soil biodiversity: The diversity of life commonly found in a soil matrix—worms, fungi, nematodes, bacteria, insects, and microcrustaceans.

Soil map: A map showing the distribution of soil types.

Soil structure: The arrangement of soil particles into units called peds. Peds are characterized by shape (blocky, columnar, granules).

Stand enhancement: Increasing the diversity of an established stand of mostly grasses by interseeding new plant species into it.

Stratification: A process whereby seeds are placed in a moist medium at appropriate temperatures for a period of time to mimic the critical conditions necessary for germination.

Stratified sampling: The designation of discrete units, such as different habitat types, within a planting site that are sampled and analyzed separately.

Succession: More or less predictable and orderly changes in the composition or structure of an ecological community over time.

Swale: A depressed area on the landscape, a wet or marshy site that can be a natural landscape or a human-created landscape to manage storm water runoff.

Tackifier: A plant-based or polymeric emulsion blend that acts as a glue to hold fiber mulch in place over the seedbed until the seed has sprouted and become established.

Tallgrass prairie: A large area of level or rolling grassland in the Mississippi River valley that in its natural, uncultivated state usually has deep, fertile soil; a cover of tall, coarse grasses; and few trees.

Thatch: A layer of dead leaves and stems on top of the soil surface.

Tillage: Disruption of the soil surface by plowing, disking, rototilling, or harrowing.

Tiller: A basal or subterranean shoot that is more or less erect and is usually associated with grasses, sedges, and rushes.

Top-kill: A condition that occurs when the above-ground parts of a plant are killed, but the remainder of the plant survives.

Topographic map: A large-scale map showing relief and human-made features of a portion of a land surface, distinguished by the portrayal of position, relation, size, shape, and elevation of the features.

Topography: The surface shapes and features of a region, including hills, valleys, lakes, streams, and other landform features.

Trash plow: A circular disk on a no-till seed drill that moves thatch away from the planting wheels.

Ungulates: Hoofed mammals such as horses, cattle, bison, deer, swine, and elephants.

Vegetative sampling: The measurement of plant characteristics within a designated sample area.

Viability: The ability of a seed to germinate.

Vigor: The ability of a plant to successfully establish in the field.

Vining: Growing in a manner that twists or entwines.

Warm-season grass: A grass that grows during the summer, then flowers and sets seed in late summer or early fall.

Weed: An undesired plant.

Wetland: Land (such as a marsh or swamp) that is covered, often intermittently, with shallow water or has soil saturated with moisture.

Winnowing: Using horizontally moving air to separate heavy from light particles.

Woodland: An area dominated by trees that reduce sunlight by 70 percent or more to the soil surface during the growing season.

Woody species: Trees and shrubs.

Xeric soil: Excessively drained and somewhat excessively drained sandy or gravelly soil and shallow loam soil on steep slopes and ridges.

"Yellow tag" seed: Native seed that is source-identified in accordance with standards set by the Association of Official Seed Certifying Agencies.

Common and Scientific Names of Plants Mentioned in This Guide

We have used *The Vascular Plants of Iowa: An Annotated Checklist and Natural History* by Lawrence J. Eilers and Dean M. Roosa (1994), *An Illustrated Guide to Iowa Prairie Plants* by Paul Christiansen and Mark Müller (1999), and the USDA Natural Resources Conservation Service PLANTS Database at http://plants.usda.gov for species selection and binomial nomenclature. Where the *Flora of North America North of Mexico* and *The Flora of Nebraska* by Robert B. Kaul, David Sutherland, and Steven Rolfsmeier (2007) provide updated nomenclature, we have included the older names in brackets below.

COMMON NAME	SCIENTIFIC NAME
Alexanders, golden	*Zizia aurea*
Alexanders, heartleaf	*Zizia aptera*
Alumroot	*Heuchera richardsonii*
Anemone, Canada	*Anemone canadensis*
Ash, green	*Fraxinus pennsylvanica*
Aspen, quaking	*Populus tremuloides*
Aster, heath	*Symphyotrichum ericoides* [*Aster ericoides*]
Aster, New England	*Symphyotrichum novae-angliae* [*Aster novae-angliae*]
Aster, silky	*Symphyotrichum sericeum* [*Aster sericeus*]
Aster, sky-blue	*Symphyotrichum oolentangiense* [*Aster azureus*]
Aster, smooth blue	*Symphyotrichum laeve* [*Aster laevis*]
Aster, upland white	*Oligoneuron album* [*Aster ptarmicoides*]
Aster, willowleaf	*Symphyotrichum praealtum* [*Aster praealtus*]
Beardtongue, foxglove	*Penstemon digitalis*
Beardtongue, large-flowered	*Penstemon grandiflorus*
Bellflower, marsh	*Campanula aparinoides*
Bergamot, wild	*Monarda fistulosa*
Black-eyed Susan	*Rudbeckia hirta*
Blazing star, meadow	*Liatris ligulistylis*
Blazing star, prairie	*Liatris pycnostachya*
Blazing star, rough	*Liatris aspera*
Blue-eyed grass	*Sisyrinchium campestre*
Blue flag	*Iris shrevei*
Bluegrass, Kentucky	*Poa pratensis*
Bluestem, big	*Andropogon gerardii*

Bluestem, little	*Schizachyrium scoparium*
Boneset	*Eupatorium perfoliatum*
Boneset, false	*Brickellia eupatorioides*
Boneset, tall	*Eupatorium altissimum*
Boxelder	*Acer negundo*
Brome, smooth	*Bromus inermis*
Bromegrass, Kalm's	*Bromus kalmii*
Buckbrush or snowberry	*Symphoricarpos occidentalis*
Buckthorn, common	*Rhamnus cathartica*
Bulrush, dark green	*Scirpus atrovirens*
Buttercup, prairie	*Ranunculus rhomboideus*
Camass, white	*Zigadenus elegans*
Cardinal flower	*Lobelia cardinalis*
Cedar, eastern red	*Juniperus virginiana*
Chicory	*Cichorium intybus*
Cinquefoil, tall	*Potentilla arguta*
Clover, purple prairie	*Dalea purpurea*
Clover, red	*Trifolium pratense*
Clover, round-headed bush	*Lespedeza capitata*
Clover, slender prairie bush	*Lespedeza leptostachya*
Clover, sweet	*Melilotus* spp.
Clover, white prairie	*Dalea candida*
Compass plant	*Silphium laciniatum*
Coneflower, fragrant	*Rudbeckia subtomentosa*
Coneflower, gray-headed	*Ratibida pinnata*
Coneflower, pale purple	*Echinacea pallida*
Coreopsis, prairie	*Coreopsis palmata*
Coreopsis, tall	*Coreopsis tripteris*
Cowbane	*Oxypolis rigidior*
Cress, spring	*Cardamine bulbosa*
Culver's root	*Veronicastrum virginicum*
Cup plant	*Silphium perfoliatum*
Daisy, ox-eye	*Chrysanthemum leucanthemum*
Dandelion, common	*Taraxacum officinale*
Dandelion, false	*Krigia biflora*
Dogwood, gray	*Cornus foemina*
Dogwood, rough-leaved	*Cornus drummondii*
Dropseed, prairie	*Sporobolus heterolepis*
Dropseed, rough or tall	*Sporobolus compositus*
Elm, Siberian	*Ulmus pumila*
Fescue, tall	*Schedonorus phoenix*
Fleabane, daisy	*Erigeron strigosus*

Foxtail	*Setaria* spp.
Garlic, wild	*Allium canadense*
Gentian, bottle	*Gentiana andrewsii*
Geranium, wild	*Geranium maculatum*
Goldenrod, giant	*Solidago gigantea*
Goldenrod, grass-leaved	*Euthamia graminifolia*
Goldenrod, Missouri	*Solidago missouriensis*
Goldenrod, old field	*Solidago nemoralis*
Goldenrod, Riddell's	*Oligoneuron riddellii* [*Solidago riddellii*]
Goldenrod, showy	*Solidago speciosa*
Goldenrod, stiff	*Oligoneuron rigidum* [*Solidago rigida*]
Grama, blue	*Bouteloua gracilis*
Grama, side-oats	*Bouteloua curtipendula*
Grass, bluejoint	*Calamagrostis canadensis*
Grass, buffalo	*Bouteloua dactyloides* [*Buchloe dactyloides*]
Grass, fowl manna	*Glyceria striata*
Grass, Indian	*Sorghastrum nutans*
Grass, June	*Koeleria macrantha*
Grass, pampas	*Miscanthus sacchariflorus*
Grass, porcupine	*Hesperostipa spartea* [*Stipa spartea*]
Grass, prairie cord	*Spartina pectinata*
Grass, reed canary	*Phalaris arundinacea*
Grass, Scribner's panic	*Dichanthelium oligosanthes*
Grass, sweet	*Hierochloe odorata*
Harebell	*Campanula rotundifolia*
Honeysuckle, Tartarian	*Lonicera tatarica*
Indigo, cream false	*Baptisia bracteata*
Indigo, white wild	*Baptisia alba*
Indigo bush	*Amorpha fruticosa*
Ironweed	*Vernonia fasciculata*
Ivy, poison	*Toxicodendron radicans*
Joe Pye weed	*Eupatorium maculatum*
Lambsquarters	*Chenopodium album*
Leadplant	*Amorpha canescens*
Lespedeza, sericea	*Lespedeza cuneata*
Lily, Michigan or Turk's cap	*Lilium michiganense*
Lily, prairie	*Lilium philadelphicum*
Lobelia, great blue	*Lobelia siphilitica*
Lobelia, spiked	*Lobelia spicata*
Locust, black	*Robinia pseudoacacia*
Locust, honey	*Gleditsia triacanthos*
Loosestrife, fringed	*Lysimachia ciliata*

Loosestrife, narrow-leaved	*Lysimachia quadriflora*
Loosestrife, purple	*Lythrum salicaria*
Loosestrife, winged	*Lythrum alatum*
Lousewort or wood betony	*Pedicularis canadensis*
Lousewort, swamp	*Pedicularis lanceolata*
Lupine, wild	*Lupinus perennis*
Maple, silver	*Acer saccharinum*
Marigold, marsh	*Caltha palustris*
Meadow-rue, purple	*Thalictrum dasycarpum*
Milkweed, butterfly	*Asclepias tuberosa*
Milkweed, common	*Asclepias syriaca*
Milkweed, swamp	*Asclepias incarnata*
Milkweed, whorled	*Asclepias verticillata*
Mint, dotted	*Monarda punctata*
Mountain mint, common	*Pycnanthemum virginianum*
Mountain mint, hairy	*Pycnanthemum pilosum*
Mountain mint, slender	*Pycnanthemum tenuifolium*
Mullein, common	*Verbascum thapsus*
New Jersey tea	*Ceanothus americanus*
Onion, nodding wild	*Allium cernuum*
Onion, wild prairie	*Allium stellatum*
Orchid, eastern prairie fringed	*Platanthera leucophaea*
Orchid, western prairie fringed	*Platanthera praeclara*
Paintbrush, Indian	*Castilleja coccinea*
Parsnip, wild	*Pastinaca sativa*
Pea, partridge	*Chamaecrista fasciculata* [*Cassia fasciculata*]
Pea, veiny	*Lathyrus venosus*
Petunia, wild	*Ruellia humilis*
Phlox, marsh	*Phlox maculata*
Phlox, prairie	*Phlox pilosa*
Plantain, prairie Indian	*Arnoglossum plantagineum* [*Cacalia plantaginea*]
Prairie smoke	*Geum triflorum*
Puccoon, hoary	*Lithospermum canescens*
Pussytoes	*Antennaria neglecta*
Quackgrass	*Elymus repens* [*Agropyron repens*]
Queen Anne's lace	*Daucus carota*
Quinine, wild	*Parthenium integrifolium*
Ragweed, common	*Ambrosia artemisiifolia*
Ragweed, great	*Ambrosia trifida*
Ragwort, golden	*Senecio aureus*
Rattlesnake master	*Eryngium yuccifolium*
Rattlesnake-root	*Prenanthes racemosa*

Redtop	*Agrostis gigantea*
Reedgrass, woodland	*Cinna arundinacea*
Rose, multiflora	*Rosa multiflora*
Rose, pasture	*Rosa carolina*
Rose, prairie	*Rosa arkansana*
Rose, wild	*Rosa* spp.
Rosinweed	*Silphium integrifolium*
Rye, Canada wild	*Elymus canadensis*
Rye, perennial	*Lolium perenne*
Rye, Virginia wild	*Elymus virginicus*
Sage, white	*Artemisia ludoviciana*
Saxifrage, swamp	*Saxifraga pensylvanica*
Sedge, awlfruit	*Carex stipata*
Sedge, Bebb's	*Carex bebbii*
Sedge, Bicknell's	*Carex bicknellii*
Sedge, broom	*Carex scoparia*
Sedge, brown fox	*Carex vulpinoidea*
Sedge, Crawe's	*Carex crawei*
Sedge, heavy	*Carex gravida*
Sedge, hummock	*Carex stricta*
Sedge, Mead's	*Carex meadii*
Sedge, oval-leaf	*Carex cephalophora*
Sedge, prairie star	*Carex interior*
Sedge, shortbeak	*Carex brevior*
Sedge, troublesome	*Carex molesta*
Sedge, woolly	*Carex pellita*
Sedge, yellow fox	*Carex annectens*
Self heal	*Prunella vulgaris* var. *lanceolata*
Shooting star	*Dodecatheon meadia*
Sneezeweed	*Helenium autumnale*
Sorrel, violet wood	*Oxalis violacea*
Spiderwort, Ohio	*Tradescantia ohiensis*
Spiderwort, prairie	*Tradescantia bracteata*
Spurge, flowering	*Euphorbia corollata*
Spurge, leafy	*Euphorbia esula*
Stargrass, yellow	*Hypoxis hirsuta*
St. John's wort, great	*Hypericum ascyron* [*Hypericum pyramidatum*]
Sumac, smooth	*Rhus glabra*
Sunflower, Maximilian	*Helianthus maximiliani*
Sunflower, ox-eye	*Heliopsis helianthoides*
Sunflower, prairie	*Helianthus pauciflorus*
Sunflower, saw-tooth	*Helianthus grosseserratus*

Sunflower, western	*Helianthus occidentalis*
Switchgrass	*Panicum virgatum*
Teasel, common	*Dipsacus sylvestris*
Teasel, cut-leaved	*Dipsacus laciniatus*
Thimbleweed	*Anemone cylindrica*
Thimbleweed, tall	*Anemone virginiana*
Thistle, bull	*Cirsium vulgare*
Thistle, Canada	*Cirsium arvense*
Thistle, musk	*Carduus nutans*
Tick trefoil, Illinois	*Desmodium illinoense*
Tick trefoil, showy	*Desmodium canadense*
Timothy	*Phleum pratense*
Timothy, upland wild	*Muhlenbergia racemosa*
Toadflax, bastard	*Comandra umbellata*
Trefoil, bird's-foot	*Lotus corniculatus*
Vervain, blue	*Verbena hastata*
Vervain, hoary	*Verbena stricta*
Vetch, American	*Vicia americana*
Vetch, crown	*Securigera varia*
Vetch, milk	*Astragalus canadensis*
Vetchling, marsh	*Lathyrus palustris*
Violet, prairie	*Viola pedatifida*
Wedgegrass, prairie	*Sphenopholis obtusata*
Wheatgrass, slender	*Agropyron trachycaulum*
Wheatgrass, western	*Pascopyrum smithii* [*Agropyron smithii*]
Woundwort	*Stachys palustris*

SCIENTIFIC NAME	COMMON NAME
Acer negundo	Boxelder
Acer saccharinum	Silver maple
Agropyron trachycaulum	Slender wheatgrass
Agrostis gigantea	Redtop
Allium canadense	Wild garlic
Allium cernuum	Nodding wild onion
Allium stellatum	Wild prairie onion
Ambrosia artemisiifolia	Common ragweed
Ambrosia trifida	Great ragweed
Amorpha canescens	Leadplant
Amorpha fruticosa	Indigo bush
Andropogon gerardii	Big bluestem
Anemone canadensis	Canada anemone
Anemone cylindrica	Thimbleweed

Anemone virginiana	Tall thimbleweed
Antennaria neglecta	Pussytoes
Arnoglossum plantagineum [*Cacalia plantaginea*]	Prairie Indian plantain
Artemisia ludoviciana	White sage
Asclepias incarnata	Swamp milkweed
Asclepias syriaca	Common milkweed
Asclepias tuberosa	Butterfly milkweed
Asclepias verticillata	Whorled milkweed
Astragalus canadensis	Milk vetch
Baptisia alba	White wild indigo
Baptisia bracteata	Cream false indigo
Bouteloua curtipendula	Side-oats grama
Bouteloua dactyloides [*Buchloe dactyloides*]	Buffalo grass
Bouteloua gracilis	Blue grama
Brickellia eupatorioides	False boneset
Bromus inermis	Smooth brome
Bromus kalmii	Kalm's bromegrass
Calamagrostis canadensis	Bluejoint grass
Caltha palustris	Marsh marigold
Campanula aparinoides	Marsh bellflower
Campanula rotundifolia	Harebell
Cardamine bulbosa	Spring cress
Carduus nutans	Musk thistle
Carex annectens	Yellow fox sedge
Carex bebbii	Bebb's sedge
Carex bicknellii	Bicknell's sedge
Carex brevior	Shortbeak sedge
Carex cephalophora	Oval-leaf sedge
Carex crawei	Crawe's sedge
Carex gravida	Heavy sedge
Carex interior	Prairie star sedge
Carex meadii	Mead's sedge
Carex molesta	Troublesome sedge
Carex pellita	Woolly sedge
Carex scoparia	Broom sedge
Carex stipata	Awlfruit sedge
Carex stricta	Hummock sedge
Carex vulpinoidea	Brown fox sedge
Castilleja coccinea	Indian paintbrush
Ceanothus americanus	New Jersey tea
Chamaecrista fasciculata [*Cassia fasciculata*]	Partridge pea
Chenopodium album	Lambsquarters

Chrysanthemum leucanthemum	Ox-eye daisy
Cichorium intybus	Chicory
Cinna arundinacea	Woodland reedgrass
Cirsium arvense	Canada thistle
Cirsium vulgare	Bull thistle
Comandra umbellata	Bastard toadflax
Coreopsis palmata	Prairie coreopsis
Coreopsis tripteris	Tall coreopsis
Cornus drummondii	Rough-leaved dogwood
Cornus foemina	Gray dogwood
Dalea candida	White prairie clover
Dalea purpurea	Purple prairie clover
Daucus carota	Queen Anne's lace
Desmodium canadense	Showy tick trefoil
Desmodium illinoense	Illinois tick trefoil
Dichanthelium oligosanthes	Scribner's panic grass
Dipsacus laciniatus	Cut-leaved teasel
Dipsacus sylvestris	Common teasel
Dodecatheon meadia	Shooting star
Echinacea pallida	Pale purple coneflower
Elymus canadensis	Canada wild rye
Elymus repens [*Agropyron repens*]	Quackgrass
Elymus virginicus	Virginia wild rye
Erigeron strigosus	Daisy fleabane
Eryngium yuccifolium	Rattlesnake master
Eupatorium altissimum	Tall boneset
Eupatorium maculatum	Joe Pye weed
Eupatorium perfoliatum	Boneset
Euphorbia corollata	Flowering spurge
Euphorbia esula	Leafy spurge
Euthamia graminifolia	Grass-leaved goldenrod
Fraxinus pennsylvanica	Green ash
Gentiana andrewsii	Bottle gentian
Geranium maculatum	Wild geranium
Geum triflorum	Prairie smoke
Gleditsia triacanthos	Honey locust
Glyceria striata	Fowl manna grass
Helenium autumnale	Sneezeweed
Helianthus grosseserratus	Saw-tooth sunflower
Helianthus maximiliani	Maximilian sunflower
Helianthus occidentalis	Western sunflower
Helianthus pauciflorus	Prairie sunflower

Heliopsis helianthoides	Ox-eye sunflower
Hesperostipa spartea [*Stipa spartea*]	Porcupine grass
Heuchera richardsonii	Alumroot
Hierochloe odorata	Sweet grass
Hypericum ascyron [*Hypericum pyramidatum*]	Great St. John's wort
Hypoxis hirsuta	Yellow stargrass
Iris shrevei	Blue flag
Juniperus virginiana	Eastern red cedar
Koeleria macrantha	June grass
Krigia biflora	False dandelion
Lathyrus palustris	Marsh vetchling
Lathyrus venosus	Veiny pea
Lespedeza capitata	Round-headed bush clover
Lespedeza cuneata	Sericea lespedeza
Lespedeza leptostachya	Slender prairie bush clover
Liatris aspera	Rough blazing star
Liatris ligulistylis	Meadow blazing star
Liatris pycnostachya	Prairie blazing star
Lilium michiganense	Michigan or Turk's cap lily
Lilium philadelphicum	Prairie lily
Lithospermum canescens	Hoary puccoon
Lobelia cardinalis	Cardinal flower
Lobelia siphilitica	Great blue lobelia
Lobelia spicata	Spiked lobelia
Lolium perenne	Perennial rye
Lonicera tatarica	Tartarian honeysuckle
Lotus corniculatus	Bird's-foot trefoil
Lupinus perennis	Wild lupine
Lysimachia ciliata	Fringed loosestrife
Lysimachia quadriflora	Narrow-leaved loosestrife
Lythrum alatum	Winged loosestrife
Lythrum salicaria	Purple loosestrife
Melilotus spp.	Sweet clover
Miscanthus sacchariflorus	Pampas grass
Monarda fistulosa	Wild bergamot
Monarda punctata	Dotted mint
Muhlenbergia racemosa	Upland wild timothy
Oligoneuron album [*Aster ptarmicoides*]	Upland white aster
Oligoneuron riddellii [*Solidago riddellii*]	Riddell's goldenrod
Oligoneuron rigidum [*Solidago rigida*]	Stiff goldenrod
Oxalis violacea	Violet wood sorrel
Oxypolis rigidior	Cowbane

Panicum virgatum	Switchgrass
Parthenium integrifolium	Wild quinine
Pascopyrum smithii [*Agropyron smithii*]	Western wheatgrass
Pastinaca sativa	Wild parsnip
Pedicularis canadensis	Lousewort or wood betony
Pedicularis lanceolata	Swamp lousewort
Penstemon digitalis	Foxglove beardtongue
Penstemon grandiflorus	Large-flowered beardtongue
Phalaris arundinacea	Reed canary grass
Phleum pratense	Timothy
Phlox maculata	Marsh phlox
Phlox pilosa	Prairie phlox
Platanthera leucophaea	Eastern prairie fringed orchid
Platanthera praeclara	Western prairie fringed orchid
Poa pratensis	Kentucky bluegrass
Populus tremuloides	Quaking aspen
Potentilla arguta	Tall cinquefoil
Prenanthes racemosa	Rattlesnake-root
Prunella vulgaris var. *lanceolata*	Self heal
Pycnanthemum pilosum	Hairy mountain mint
Pycnanthemum tenuifolium	Slender mountain mint
Pycnanthemum virginianum	Common mountain mint
Ranunculus rhomboideus	Prairie buttercup
Ratibida pinnata	Gray-headed coneflower
Rhamnus cathartica	Common buckthorn
Rhus glabra	Smooth sumac
Robinia pseudoacacia	Black locust
Rosa arkansana	Prairie rose
Rosa carolina	Pasture rose
Rosa multiflora	Multiflora rose
Rosa spp.	Wild rose
Rudbeckia hirta	Black-eyed Susan
Rudbeckia subtomentosa	Fragrant coneflower
Ruellia humilis	Wild petunia
Saxifraga pensylvanica	Swamp saxifrage
Schedonorus phoenix	Tall fescue
Schizachyrium scoparium	Little bluestem
Scirpus atrovirens	Dark green bulrush
Securigera varia	Crown vetch
Senecio aureus	Golden ragwort
Setaria spp.	Foxtail
Silphium integrifolium	Rosinweed

Silphium laciniatum	Compass plant
Silphium perfoliatum	Cup plant
Sisyrinchium campestre	Blue-eyed grass
Solidago gigantea	Giant goldenrod
Solidago missouriensis	Missouri goldenrod
Solidago nemoralis	Old field goldenrod
Solidago speciosa	Showy goldenrod
Sorghastrum nutans	Indian grass
Spartina pectinata	Prairie cord grass
Sphenopholis obtusata	Prairie wedgegrass
Sporobolus compositus	Rough or tall dropseed
Sporobolus heterolepis	Prairie dropseed
Stachys palustris	Woundwort
Symphoricarpos occidentalis	Buckbrush or snowberry
Symphyotrichum ericoides [*Aster ericoides*]	Heath aster
Symphyotrichum laeve [*Aster laevis*]	Smooth blue aster
Symphyotrichum novae-angliae [*Aster novae-angliae*]	New England aster
Symphyotrichum oolentangiense [*Aster azureus*]	Sky-blue aster
Symphyotrichum praealtum [*Aster praealtus*]	Willowleaf aster
Symphyotrichum sericeum [*Aster sericeus*]	Silky aster
Taraxacum officinale	Common dandelion
Thalictrum dasycarpum	Purple meadow-rue
Toxicodendron radicans	Poison ivy
Tradescantia bracteata	Prairie spiderwort
Tradescantia ohiensis	Ohio spiderwort
Trifolium pratense	Red clover
Ulmus pumila	Siberian elm
Verbascum thapsus	Common mullein
Verbena hastata	Blue vervain
Verbena stricta	Hoary vervain
Vernonia fasciculata	Ironweed
Veronicastrum virginicum	Culver's root
Vicia americana	American vetch
Viola pedatifida	Prairie violet
Zigadenus elegans	White camass
Zizia aurea	Golden alexanders
Zizia aptera	Heartleaf alexanders

References

Anderson, P. 1996. *GIS research to digitize maps of Iowa 1832–1859: Vegetation from general land office township plat maps.* Ames: Living Roadway Trust Fund, Iowa Department of Transportation.

Barnhart, S. 2002. *Improving pasture by frost seeding.* Ames: Iowa State University Extension.

Baskin, C. C., and J. M. Baskin. 1998. *Seeds: Ecology, biogeography, and evolution of dormancy and germination.* San Diego: Academic Press.

Biondini, M. E., A. A. Steuter, and C. E. Grygiel. 1989. Seasonal fire effects on the diversity patterns, spatial distribution and community structure of forbs in the Northern Mixed Prairie. *Plant Ecology* 85:21–31.

Bragg, T. B. 1978. Allwine: Prairie preserve: A reestablished bluestem grassland research area. In *Proceedings of the fifth Midwest Prairie Conference, August 22–24, 1976.* Ames: Iowa State University.

———. 1982. Seasonal variations in fuel and fuel consumption by fires in a bluestem prairie. *Ecology* 63:7–11.

——— and L. C. Hulbert. 1976. Woody plant invasion of unburned Kansas bluestem prairie. *Journal of Range Management* 29:19–23.

Briggs, J. M., and D. J. Gibson. 1992. Effect of burning on tree spatial patterns in a tallgrass prairie landscape. *Bulletin of the Torrey Botanical Club* 119:300–307.

———, A. K. Knapp, and B. L. Block. 2002. Expansion of woody plants in tallgrass prairie: A 15-year study of fire and fire-grazing interactions. *American Midland Naturalist* 147:287–294.

———, ———, J. M. Blair, J. L. Heisler, G. A. Hoch, M. S. Lett, and J. K. McCarron. 2005. An ecosystem in transition: Causes and consequences of the conversion of mesic grassland to shrubland. *BioScience* 55(3):243–254.

Butler, J. L., and D. R. Cogan. 2004. Leafy spurge effects on patterns of plant species richness. *Journal of Range Management* 57(3):305–311.

Callicott, J. B., and E. T. Freyfogle. 1999. *For the health of the land: Previously unpublished essays and other writings.* Washington, D.C.: Island Press.

Cawley, M. 1986. The structure of plant communities. In *Plant ecology,* ed. M. Cawley. Oxford: Blackwell Publications.

Chadde, S. W. 2002. *Great Lakes wetland flora.* 2d ed. Laurium, Mich.: Pocket Flora Press.

Christiansen, P., and M. Müller. 1999. *An illustrated guide to Iowa prairie plants.* Iowa City: University of Iowa Press.

Clark, O. R. 1937. Interception of rainfall by herbaceous vegetation. *Science* 86:591–592.

Clements, F. E. 1928. *Plant succession and indicators*. New York: H. W. Wilson.

Clewell, A., J. Rieger, and J. Monro. 2005. *Guidelines for developing and managing ecological restoration projects.* 2d ed. Tucson: Society for Ecological Restoration International.

Cohen, D. 2007. Personal communication.

Cohen, J. 1998. The impacts of fire on ecosystems. Austin: University of Texas. http://www.micro.utexas.edu/courses/mcmurry/spring98/10/jerrycohen.

Collins, S. L. 2000. Disturbance frequency and community stability in native tallgrass prairie. *American Naturalist* 155:311–325.

Conard, E. C. 1954. Effect of time of cutting on yield and botanical composition of prairie hay in southeastern Nebraska. *Journal of Range Management* 7(4):181–182.

Cooper, H. W. 1957. Some plant materials and improved techniques used in soil and water conservation in the Great Plains. *Journal of Soil and Water Conservation* 12:163–168.

Crop Data Management Systems. 2005. http:www.cdms.net.

Davison, C., and K. Kindscher. 1999. Tools for diversity: Fire, grazing and mowing on tallgrass prairies. *Ecological Restoration* 17(3):136–143.

Dayton, R. 1988. Guidelines for herbaceous stand evaluation. Iowa Agronomy Technical Note 19. Des Moines: United States Department of Agriculture, Natural Resources Conservation Service, Iowa State Office.

Deno, N. C. 1993. *Seed germination theory and practice: Based on experiments on 145 families, 805 genera, and about 4,000 species.* 2d ed. Self-published.

Diamond, J. M. 1975. The island dilemma: Lessons of modern biogeographic studies for the designing of natural reserves. *Biological Conservation* 7:129–146.

Ecological Restoration International, Science and Policy Working Group. 2004. *The SER International primer on ecological restoration.* Tucson: Society for Ecological Restoration International.

Engstrom, C. 2004. Return of the prairie. *Iowa Natural Heritage* (Fall).

Ewing, A. L., and D. M. Engle. 1988. Effects of late summer fire on tallgrass prairie microclimate and community composition. *American Midland Naturalist* 120:212–223.

Foster, L. 2004. Personal communication.

Fuhlendorf, S. D., and D. M. Engle. 2001. Restoring heterogeneity on rangelands: Ecosystem management based on evolutionary grazing patterns. *Bioscience* 51:625–632.

——— and ———. 2004. Application of the fire-grazing interaction to restore a shifting mosaic on tallgrass prairie. *Journal of Applied Ecology* 41:604–614.

Gibson, D. J. 1989. Hulbert's study of factors effecting botanical composition of tallgrass prairie. In *Proceedings of the eleventh North American Prairie Conference, August 7–11, 1988.* Lincoln: University of Nebraska Press.

Gilpin, M. E., and M. E. Soule. 1986. Minimum viable populations: Process of species extinctions. In *Conservation biology: The science of scarcity and diversity,* ed. M. E. Soule. Sunderland, Mass.: Sinauer.

Global Invasive Species Team. Invasive species, control methods, and best management practices. http://www.invasive.org/gist/control.html.

———. What are the worst weeds? http://www.invasive.org/gist/worst.html.

Harrington, J. F. 1972. Seed storage and longevity. In *Seed biology*, vol. 3, ed. T. T. Kazlowski. New York: Academic Press.

Heisler, J. L., J. M. Briggs, and A. K. Knapp. 2003. Long-term patterns of shrub expansion in a C_4-dominated grassland: Fire frequency and the dynamics of shrub cover and abundance. *American Journal of Botany* 90:423–428.

Helzer, C. J. 2001. Managing prairies for biodiversity. *Platte River Current* newsletter. Wood River, Neb.: Crane Meadows Nature Center.

——— and A. A. Steuter. 2005. Preliminary effects of patch-burn grazing on a high-density prairie restoration. *Ecological Restoration* 23(3):167–171.

Hemsath, C. 2007. Quantifying granivory in a reconstructed prairie: Effects of season, species, seed predators, sacrificial food, and the chemical deterrent capsaicin. Master's thesis, University of Northern Iowa.

Henderson, K., and C. Kern. 1999. Integrated roadside vegetation management technical manual. Cedar Falls: Roadside Management Program, University of Northern Iowa.

Henderson, R. A. 1992. Ten-year response of a Wisconsin prairie remnant to seasonal timing of fire. In *Proceedings of the twelfth North American Prairie Conference, August 5–9, 1990*. Cedar Falls: University of Northern Iowa.

Hessel, S. A. 1954. A guide to collecting plant-boring larvae of the genus *Papaipema* (Noctuidae). *Lepidopterists' News* 8:57–63.

Hough, M. 2004. *Cities and natural processes: A basis for sustainability*. New York: Routledge, Taylor and Francis Group.

Howe, H. F. 1994. Managing species diversity in tallgrass prairie: Assumptions and implications. *Conservation Biology* 8(3):691–704.

———. 1999a. Dominance, diversity and grazing in tallgrass restoration. *Ecological Restoration* 17(1 and 2):59–66.

———. 1999b. Response of *Zizea aurea* to seasonal mowing and fire in a restored prairie. *American Midland Naturalist* 141:373–380.

Hulbert, L. C. 1986. Fire effects on tallgrass prairie. In *Proceedings of the ninth North American Prairie Conference, July 29–August 1, 1984*. Fargo: Tri-College University Center for Environmental Studies.

Illinois Natural History Survey. INHS botany collection. http://www.inhs.uiuc.edu/cbd/collections/botany/botanyintro.html.

Illinois Plant Information Network. http://www.fs.fed.us/ne/delaware/ilpin/A.htm.

Illinois Wildflowers. http://www.illinoiswildflowers.info/.

Iowa State University Department of Ecology, Evolution, and Organismal Biology. Grasses of Iowa. http://www.eeob.iastate.edu/research/iowagrasses/index.html.

Jackson, L., and L. Dittmer. 1997. *Prairie seedlings illustrated: An identification guide*. Cedar Falls: University of Northern Iowa.

Jacobson, E. T., D. A. Tober, R. J. Haas, and D. C. Darris. 1986. The performance of selected cultivars of warm season grasses in the northern prairie and plains states. In *Proceedings of the ninth North American Prairie Conference, July 29–August 1, 1984*. Fargo: Tri-College University Center for Environmental Studies.

Jameson, E. W. 1947. Natural history of the prairie vole (Mammalian genus *Microtus*). Lawrence: University of Kansas Publications, Museum of Natural History.

Jog, S. K., E. J. Questad, K. Kindscher, B. L. Foster, and H. Loring. 2006. Floristic quality as an indicator of native species diversity within managed grasslands. *Natural Areas Journal* 26:149–167.

Kansas Wildflowers and Grasses. http://www.lib.ksu.edu/wildflower/.

Kaspar, T., B. Knutson, K. Kohler, and J. Singer. 2004. Information summary prepared for Iowa NRCS. Ames: USDA-ARS National Soil Tilth Laboratory.

Khadduri, N. Y., and J. T. Harrington. 2002. Shaken, not stirred: A percussion scarification technique. *Native Plants Journal* (Spring).

Kilde, R., and E. Fuge. 2002. *Going native: A prairie restoration handbook for Minnesota landowners*. St. Paul: Minnesota Department of Natural Resources.

Kline, V. M. 1986. Response of sweet clover *(Melilotus alba* Desr.) and associated prairie vegetation to seven experimental burning and mowing treatments. In *Proceedings of the ninth North American Prairie Conference, July 29–August 1, 1984*. Fargo: Tri-College University Center for Environmental Studies.

——— and E. A. Howell. 1987. Prairies. In *Restoration ecology: A synthetic approach to ecological research*, ed. W. Jordan, W. M. Gilpin, and J. Aber. Cambridge: Cambridge University Press.

Knapp, A. K., J. R. Briggs, D. C. Hartnett, and D. W. Kaufman. 1992. Long term ecological research at the Konza Prairie Research Natural Area. Site description and research summary (1981–91). Manhattan, Kans.

Knapp, E. E., and K. J. Rice. 1996. Genetic structure and gene flow in *Elymus glaucus* (blue wildrye): Implications for native grassland restoration. *Restoration Ecology* 4:1–10.

Kucera, C. L. 1960. Forest encroachment in native prairie. *Iowa State Journal of Science* 34(4):635–639.

———. 1970. Ecological effects of fire on tallgrass prairie. In *Symposium on prairie and prairie restoration: Proceedings of the first Midwest Prairie Conference, September 14–15, 1968*. Galesburg, Ill.: Knox College.

———. 1990. Fire studies at Tucker Prairie. *Iowa Prairie Blazing Star* 2(1):3–6.

Kurtz, C. 2001. *A practical guide to prairie reconstruction*. Iowa City: University of Iowa Press.

Ladd, D., and F. Oberle. 1995. *Tallgrass prairie wildflowers*. Helena-Billings: Falcon Press.

Lady Bird Johnson Wildflower Center. Native plant information network. http://www.wildflower.org/explore/.

Lalonde, G., and B. Roitberg. 1994. Mating system, life-history, and reproduction in Canada thistle (*Cirsium arvense*; Asteraceae). *American Journal of Botany* 81:21–28.

Lekwa, S. 1984. Prairie restoration and management. *Iowa Conservationist* 43(9): 12–14.

Leopold, A. 1934. The arboretum and the university. *Parks and Recreation* 18:59–60.

———. 1949. *Sand county almanac.* Oxford: Oxford University Press.

Lindroth, R. L., and G. O. Batzli. 1984. Food habits of the meadow vole (*Microtus pennsylvanicus*) in bluegrass and prairie habitats. *Journal of Mammalogy* 65:600–606.

Maag, S. 1994. Perceptions of prairie in corporate settings: A study in adaptation. In *Prairie biodiversity: Proceedings of the fourteenth North American Prairie Conference, July 12–16, 1994.* Manhattan: Kansas State University.

Madson, J. 1972. The running country. *Audubon* 74(4):4–19.

Mariner, R., D. Dreher, K. Huebner, L. Hill, and E. Wurm. 1997. Source book on natural landscaping for local officials. Northeastern Illinois Planning Commission.

McGovern, J. 2007. Personal communication.

McMillan, C. 1959. The role of ecotypic variation in the distribution of the central grassland of North America. *Ecological Monographs* 29:285–308.

Menges, E. S. 1991. Seed germination percentage increases with population size in a fragmented prairie species. *Conservation Biology* 5:158–164.

Meyer, M. H., and V. A. Gaynor. 2002. Effect of seeding dates on establishment of native grasses. *Native Plants Journal* 3(2):132–138.

Meyermann, P. 2006, 2008. Personal communication.

Miller, S. G., S. P. Bratton, and J. Hadidian. 1992. Impacts of white-tailed deer on endangered and threatened vascular plants. *Natural Areas Journal* 12:67–74.

Missouri Department of Conservation. Go native! Plant identification. http://mdc.mo.gov/grownative/plantID/.

———. 2004. Managing Missouri's hay prairies. http://mdc.mo.gov/landown/grass/hay/.

Moats, S. 2006. Personal communication.

Morgan, J. 1997. Plowing and seeding. In *The tallgrass restoration handbook: For prairies, savannas, and woodlands*, ed. S. Packard and C. Mutel. Washington, D.C.: Island Press.

———, D. Collicut, and J. Durant. 1995. *Restoring Canada's native prairies: A practical manual.* Argyle, Manitoba: Prairie Habitats.

Nagel, H. G. 1983. Effect of spring burning date on mixed-prairie soil moisture, productivity, and plant species composition. In *Proceedings of the seventh North American Prairie Conference, August 4–6, 1980.* Springfield: Southwestern Missouri State University.

Northern Prairie Wildlife Research Center. Aquatic and wetland vascular plants of the northern Great Plains. http://www.npwrc.usgs.gov/resource/1999/vascplnt/vascplnt.htm#contents.

———. Midwestern wetland flora. http://www.npwrc.usgs.gov/resource/plants/floramw/.

Noss, R. T., R. T. LaRoe III, and J. M. Scott. 1995. Endangered ecosystems of the United States. *Biological Report* 28:83.

Nyboer, R. W. 1989. Grazing as a factor in the decline of Illinois hill prairies. In *Proceedings of the sixth North American Prairie Conference, August 12–17, 1978*. Columbus: Ohio State University.

Olson, W. W. 1986. Phenology of selected varieties of warm season native grasses. In *Proceedings of the ninth North American Prairie Conference, July 29–August 1, 1984*. Fargo: Tri-College University Center for Environmental Studies.

Opler, P. A. 1981. Management of prairie habitats for insect conservation. *Journal of the Natural Areas Association* 1(4):3–6.

Packard, S., and C. Mutel. 1997, 2005. *The tallgrass restoration handbook: For prairies, savannas, and woodlands*. Washington, D.C.: Island Press.

Panzer, R. 1988. Managing prairie remnants for insect conservation. *Natural Areas Journal* 8(2):83–90.

——— and M. Schwartz. 2000. Effects of management burning on prairie insect species richness within a system of small, highly fragmented reserves. *Biological Conservation* 96:363–369.

Peterson, R. T., and M. McKenney. 1998. *Field guide to wildflowers: Northeastern and north-central North America*. New York: Houghton Mifflin.

Pohl, R. W. 1978. *How to know the grasses*. Chicago: Lakeside Press.

Prairie Wildflowers of Illinois. http://www.illinoiswildflowers.info/prairie/plant_index.htm.

Reed, C. C. 2004. Keeping invasive plants out of restorations. *Ecological Restoration* 22(3):210–216.

Reinartz, J. R. 1997. Restoring populations of rare plants. In *The tallgrass restoration handbook: For prairies, savannas, and woodlands*, ed. S. Packard and C. Mutel. Washington, D.C.: Island Press.

Riechert, S. E., and W. G. Reeder. 1970. Effects of fire on spider distribution in southwestern Wisconsin prairies. In *Proceedings of the second Midwest Prairie Conference, September 18–20, 1970*. Madison: University of Wisconsin.

Ries, L., D. Debinski, and M. Wieland. 2001. Conservation value of roadside prairie restoration to butterfly communities. *Conservation Biology* 15:401–411.

Riley, J. 1957. Soil temperatures as related to corn yield in Iowa. *Monthly Weather Review* 85(12):393–400.

Ritterbusch, C. 2007. Curtis prairie restoration. www.prairieworksinc.com/2007/08/28/curtis-prairie-restoration.

Robertson, K. R., R. C. Anderson, and M. W. Schwartz. 1997. The tallgrass prairie mosaic. In *Conservation of highly fragmented landscapes*, ed. M. Schwartz. New York: Chapman and Hall.

Rock, H. 1972. *Prairie propogation handbook*. Milwaukee County Park, Hales Corners, Wisconsin.

Rosburg, T. 2006. Personal communication.

Runkel, S., and D. Roosa. 2009. *Wildflowers of the tallgrass prairie: The Upper Midwest*. 2d ed. Iowa City: University of Iowa Press.

Sample, D. W., and M. J. Mossman. 1997. *Managing habitat for grassland birds*. Madison: Wisconsin Department of Natural Resources.

Samuel Roberts Noble Image Foundation. Noble plant image gallery. http://www.noble.org/webapps/plantimagegallery/.

Saunders, D. A., R. J. Hobbs, and C. R. Margules. 1991. Biological consequences of ecosystem fragmentation: A review. *Conservation Biology* 5:18–32.

Schramm, P. 1992. Prairie restoration: A twenty-five-year perspective on establishment and management. In *Proceedings of the twelfth North American Prairie Conference, August 5–9, 1990*. Cedar Falls: University of Northern Iowa.

Schweitzer, D. F. 1985. Effects of prescribed burning on rare Lepidoptera: Memorandum to TNC stewardship and heritage staffs, eastern and midwestern regions. Boston: The Nature Conservancy.

Sheley, R., K. Goodwin, and M. Rinella. 2001. Mowing to manage noxious weeds. Agricultural Extension Publication. Weeds A-16 (Range and Pasture). Bozeman: Montana State University.

Shirley, S. 1994. *Restoring the tallgrass prairie: An illustrated manual for Iowa and the Upper Midwest*. Iowa City: University of Iowa Press.

Siefert, T., and T. Rosburg. 2004. Ecological research on the Interstate 35 prairie reconstruction, Story County, Iowa from 1996 to 2003. Final report. Submitted to the Roadside Development Section, Iowa Department of Transportation.

Smith, D. D. 1998. Iowa prairie: Original extent and loss, preservation and recovery attempts. *Journal of the Iowa Academy of Science* 105(3):94–108.

Smith, R., and S. Smith. 1980. *The prairie garden: 70 native plants you can grow in town or country*. Madison: University of Wisconsin Press.

——— and ———, eds. 1998. Native grass seed production manual. United States Department of Agriculture, Natural Resources Conservation Service.

Sperry, T. M. 1983. Analysis of the University of Wisconsin-Madison prairie restoration project. In *Proceedings of the eighth North American Prairie Conference, August 1–4, 1982*. Springfield: Southwest Missouri State University.

Steinauer, G., B. Whitney, K. Adams, and M. Bullerman. 2003. *A guide to prairie and wetland restoration in eastern Nebraska*. Aurora: Plains Resource Institute and Nebraska Game and Parks Commission.

Texas A&M University. Cyber sedge, bioinformatics working group. http://www.csdl.tamu.edu/FLORA/carex/carexout.htm.

Tilman, D. 1997. Community invasibility, recruitment limitation, and grassland biodiversity. *Ecology* 78:81–92.

——— and J. Downing. 1994. Biodiversity and stability in grasslands. *Nature* 367: 356–363.

Towne, E. G., D. C. Hartnett, and R. C. Cochran. 2005. Vegetation trends in tallgrass prairie from bison and cattle grazing. *Ecological Applications* 15(5):1550–1559.

——— and P. D. Ohlenbusch. 1992. Rangeland brush management. Kansas State University Agricultural Experiment Station and Cooperative Extension Service.

——— and C. Owensby. 1984. Long-term effects of annual burning at different dates in ungrazed Kansas tallgrass prairie. *Journal of Range Management* 37(5):392–397.

Truax Company. 2004. FLEXII drills operator's manual.

United States Department of Agriculture, Iowa Natural Resources Conservation Service. 2002. Pasture and hay planting. Code 512.

United States Department of Agriculture, Natural Resources Conservation Service. BADLANDS ecotype little bluestem. http://www.plant-materials.nrcs.usda.gov/pubs/ndpmcrb7356.pdf/.

United States Department of Agriculture, Natural Resources Conservation Service. PLANTS Database. http://plants.usda.gov.

University of Minnesota. http://www.cedarcreek.umn.edu/herbarium/.

Weaver, J. E. 1954. *North American prairie*. Lincoln, Neb.: Johnsen Publishing Company.

West, D., and D. Undersander. 1997. Spring frost seeding. In *Proceedings of the Wisconsin forage production and use symposium*. Wisconsin Dells.

Williams, D. 2002. Emergence and mortality of native prairie forbs seeded into an established stand of grasses. Master's thesis, University of Northern Iowa.

———, G. Houseal, and D. Smith. 2006. Growth and reproduction of local ecotype and cultivated varieties of *Panicum virgatum* and *Coreopsis palmata* grown in common gardens. In *Proceedings of the nineteenth North American Prairie Conference, July 23–26, 2004*. Madison: University of Wisconsin.

———, L. Jackson, and D. Smith. 2007. Effects of frequent mowing on survival and persistence of forbs seeded into a species-poor grassland. *Restoration Ecology* 15:24–33.

Willson, G. D. 1992. Morphological characteristics of smooth brome used to determine a prescribed burn date. In *Proceedings of the twelfth North American Prairie Conference, August 5–9, 1990*. Cedar Falls: University of Northern Iowa.

——— and J. Stubbendieck. 1997. Fire effects on four growth stages of smooth brome (*Bromus inermis*). *Natural Areas Journal* 17(4):306–312.

Wisconsin State Herbarium, University of Wisconsin, Madison. Wisflora: Wisconsin vascular plant species. http://www.botany.wisc.edu/wisflora/.

Witmer, S. 1999. *Statistics for the life sciences.* Upper Saddle River, N.J.: Prentice Hall.

Wright, H. A., and A. W. Bailey. 1982. *Fire ecology: United States and southern Canada.* New York: Wiley and Sons.

Young, S. A. 1995. Verification of germplasm origin and genetic status by seed certification agencies. In *Proceedings of the wildland shrub and arid land restoration symposium, October 19–21, 1993.* Ogden, Utah: U.S. Department of Agriculture, Forest Service, Intermountain Research Station.

Index

See the tables on pages 20–21, 30–34, 40–41, 64, 66–68, 87, 90, 178–179, 213–220, 243–246, and 256–258 for information about individual species with regard to seed-test results, recommended native seed mixes, perennial plants that should be killed prior to planting prairie, sampling, seeding, germination, propagation, transplanting, and harvesting periods.

OTHER BUR OAK BOOKS OF INTEREST

The Butterflies of Iowa
By Dennis W. Schlicht, John C. Downey, and Jeffrey Nekola

A Country So Full of Game:
The Story of Wildlife in Iowa
By James J. Dinsmore

Deep Nature: Photographs from Iowa
Photographs by Linda Scarth and Robert Scarth, essay by John Pearson

The Ecology and Management of Prairies in the Central United States
By Chris Helzer

The Elemental Prairie:
Sixty Tallgrass Plants
By George Olson and John Madson

The Emerald Horizon:
The History of Nature in Iowa
By Cornelia F. Mutel

Enchanted by Prairie
By Bill Witt and Osha Gray Davidson

An Illustrated Guide to
Iowa Prairie Plants
By Paul Christiansen and Mark Müller

The Iowa Breeding Bird Atlas
By Laura Spess Jackson,
Carol A. Thompson, and
James J. Dinsmore

The Iowa Nature Calendar
By Jean C. Prior and James Sandrock,
illustrated by Claudia McGehee

Landforms of Iowa
By Jean C. Prior

A Practical Guide to
Prairie Reconstruction
By Carl Kurtz

Prairie: A North American Guide
By Suzanne Winckler

Prairie in Your Pocket:
A Guide to Plants of the Tallgrass Prairie
By Mark Müller

Prairies, Forests, and Wetlands:
The Restoration of Natural Landscape Communities in Iowa
By Janette R. Thompson

Restoring the Tallgrass Prairie:
An Illustrated Manual for Iowa and the Upper Midwest
By Shirley Shirley

A Tallgrass Prairie Alphabet
By Claudia McGehee

The Tallgrass Prairie Center Guide to Seed and Seedling Identification in the Upper Midwest
By Dave Williams,
illustrated by Brent Butler

The Vascular Plants of Iowa: An Annotated Checklist and Natural History
By Lawrence J. Eilers and Dean M. Roosa

A Watershed Year: Anatomy of the Iowa Floods of 2008
Edited by Cornelia F. Mutel

Where the Sky Began: Land of the Tallgrass Prairie
By John Madson

Wildflowers and Other Plants of Iowa Wetlands
By Sylvan T. Runkel and Dean M. Roosa

Wildflowers of Iowa Woodlands
By Sylvan T. Runkel and Alvin F. Bull

Wildflowers of the Tallgrass Prairie: The Upper Midwest
By Sylvan T. Runkel and Dean M. Roosa